LARSON
BOSWELL
STIFF

Applying • Reasoning • Measuring

Chapter 10 Resource Book

The Resource Book contains the wide variety of blackline masters available for Chapter 10. The blacklines are organized by lesson. Included are support materials for the teacher as well as practice, activities, applications, and assessment resources.

McDougal Littell
A HOUGHTON MIFFLIN COMPANY
Evanston, Illinois • Boston • Dallas

Contributing Authors

The authors wish to thank the following individuals for their contributions to the Chapter 10 Resource Book.

Eric J. Amendola
Karen Collins
Michael Downey
Patrick M. Kelly
Edward H. Kuhar
Lynn Lafferty
Dr. Frank Marzano
Wayne Nirode
Dr. Charles Redmond
Paul Ruland

ISBN: 0-618-02073-X

6789-VEI- 04 03

Contents

10 *Circles*

Contents

Contents

Descriptions of Resources

This Chapter Resource Book is organized by lessons within the chapter in order to make your planning easier. The following materials are provided:

Tips for New Teachers These teaching notes provide both new and experienced teachers with useful teaching tips for each lesson, including tips about common errors and inclusion.

Parent Guide for Student Success This guide helps parents contribute to student success by providing an overview of the chapter along with questions and activities for parents and students to work on together.

Prerequisite Skills Review Worked-out examples are provided to review the prerequisite skills highlighted on the Study Guide page at the beginning of the chapter. Additional practice is included with each worked-out example.

Strategies for Reading Mathematics The first page teaches reading strategies to be applied to the current chapter and to later chapters. The second page is a visual glossary of key vocabulary.

Lesson Plans and Lesson Plans for Block Scheduling This planning template helps teachers select the materials they will use to teach each lesson from among the variety of materials available for the lesson. The block-scheduling version provides additional information about pacing.

Warm-Up Exercises and Daily Homework Quiz The warm-ups cover prerequisite skills that help prepare students for a given lesson. The quiz assesses students on the content of the previous lesson. (Transparencies also available)

Activity Support Masters These blackline masters make it easier for students to record their work on selected activities in the Student Edition.

Alternative Lesson Openers An engaging alternative for starting each lesson is provided from among these four types: ***Application, Activity, Geometry Software,*** or ***Visual Approach.*** (Color transparencies also available)

Technology Activities with Keystrokes Keystrokes for Geometry software and calculators are provided for each Technology Activity in the Student Edition, along with alternative Technology Activities to begin selected lessons.

Practice A, B, and C These exercises offer additional practice for the material in each lesson, including application problems. There are three levels of practice for each lesson: A (basic), B (average), and C (advanced).

Contents

Reteaching with Practice These two pages provide additional instruction, worked-out examples, and practice exercises covering the key concepts and vocabulary in each lesson.

Quick Catch-Up for Absent Students This handy form makes it easy for teachers to let students who have been absent know what to do for homework and which activities or examples were covered in class.

Cooperative Learning Activities These enrichment activities apply the math taught in the lesson in an interesting way that lends itself to group work.

Interdisciplinary Applications/Real-Life Applications Students apply the mathematics covered in each lesson to solve an interesting interdisciplinary or real-life problem.

Math and History Applications This worksheet expands upon the Math and History feature in the Student Edition.

Challenge: Skills and Applications Teachers can use these exercises to enrich or extend each lesson.

Quizzes The quizzes can be used to assess student progress on two or three lessons.

Chapter Review Games and Activities This worksheet offers fun practice at the end of the chapter and provides an alternative way to review the chapter content in preparation for the Chapter Test.

Chapter Tests A, B, and C These are tests that cover the most important skills taught in the chapter. There are three levels of test: A (basic), B (average), and C (advanced).

SAT/ACT Chapter Test This test also covers the most important skills taught in the chapter, but questions are in multiple-choice and quantitative-comparison format. (See *Alternative Assessment* for multi-step problems.)

Alternative Assessment with Rubrics and Math Journal A journal exercise has students write about the mathematics in the chapter. A multi-step problem has students apply a variety of skills from the chapter and explain their reasoning. Solutions and a 4-point rubric are included.

Project with Rubric The project allows students to delve more deeply into a problem that applies the mathematics of the chapter. Teacher's notes and a 4-point rubric are included.

Cumulative Review These practice pages help students maintain skills from the current chapter and preceding chapters.

CHAPTER 10

Tips for New Teachers

For use with Chapter 10

Lesson 10.1

COMMON ERROR Secants and tangents are defined as lines on page 595. Students have a tendency to write them as segments when identifying them in relation to a circle. They may also be confused when looking at the diagrams on page 598. The statement of Theorem 10.3 uses the word "segments" because the theorem applies to the part of each secant line from the intersection of the secants to the point of tangency. Examples 6 and 7 indicate lines in the given notation, but show segments in the diagrams. Be sure students are clear about what you will accept for diagrams and answers. You may want to decide that both forms are equally acceptable.

TEACHING TIP In Example 7 on page 598, the solution for x is shown as ± 3. This may be confusing to students who are thinking that length is always positive. Stress that the solution for x is correct because $AD = 11$ when $x = -3$. Therefore both possibilities should be identified.

Lesson 10.2

TEACHING TIP Let students know that three letters are used only when naming major arcs and semicircles. They should be aware that the order of the letters also indicates the direction of movement around the circle for all arcs.

TEACHING TIP As you discuss the properties about chords of circles, remind students to use their prior knowledge about right angles, right triangles, and perhaps even auxiliary lines. Examples 6 and 7 on page 606 are good illustrations for this reminder.

Lesson 10.3

TEACHING TIP Activity 10.3 on page 612 leads nicely into the theorem for inscribed angles on page 613. Confirm that students remember that the measure of a central angle is equal to the measure of its intercepted arc.

COMMON ERROR Students get confused visually when looking at many segments in a circle at one time, as in Example 2 on page 613. To avoid errors, have them focus on one angle at a time and identify the type of angle and its intercepted arc. Have them do that for each angle in a diagram. Once the identification is made and written down, students will be ready to proceed with the problem.

TEACHING TIP Have students recall information about the sum of the measures of the interior angles of triangles and quadrilaterals prior to discussing the theorems about inscribed polygons on page 615.

Lesson 10.4

INCLUSION You might want to have students pay special attention to the figure in Theorem 10.14 in which two tangents intersect. Ask students to find $m\widehat{PQR}$ if $m\widehat{PR} = 50°$. Students with learning difficulties may need help in seeing that in this case the sum of the measures of the intercepted arcs is 360°.

Lesson 10.5

TEACHING TIP Activity 10.5 leads into the statement of Theorem 10.15 on page 629 about the products of the segments of two intersecting chords. Caution students to write the product of the parts using labels first and then substituting the given lengths.

COMMON ERROR In Example 2 on page 630, students may write the algebraic equation first. When doing that they tend to forget that the length of the whole secant segment must be used as one part of the product, along with the external segment. You may want to provide additional diagrams and lengths to give students extra practice in writing the correct product relationships and equations.

CHAPTER 10 CONTINUED

Tips for New Teachers

For use with Chapter 10

Lesson 10.6

INCLUSION Point out and illustrate the Study Tip on page 637. Show students how to sketch a circle when they know the center and the radius. The technique of marking a few points in the vertical and horizontal directions from the center will work nicely whenever the students do not have a compass to use. This technique is very easy to use when a coordinate plane is involved. Sketch the circle lightly at first to keep it rounded. Showing this example should help students with limited English proficiency better understand the process.

TEACHING TIP You may want to show students that using the standard form for the equation of a circle and substituting the coordinates of the origin, (0, 0), for (h, k) will yield the equation $x^2 + y^2 = r^2$.

Lesson 10.7

TEACHING TIP Emphasize the summary steps on page 642 for finding a locus. The steps are very important when solving a locus problem in which two or more conditions are presented. Example 3 on page 643 has several conditions stated. When the drawings are done, the result is 3 possible cases. Students need to know that each case must be drawn and explored. Also, their final answers must take each case into consideration.

Outside Resources

BOOKS/PERIODICALS

Samide, Andrew J. and Amanda M. Warfield. "A Mean Solution to an Old Circle Standard." *Mathematics Teacher* (May 1996); pp. 411–413.

Libow, Herb. "Explorations in Geometry: The 'Art' of Mathematics." *Mathematics Teacher* (May 1997); pp. 340–342.

SOFTWARE

Geometry Through the Circle with The Geometer's Sketchpad. Blackline masters and Macintosh and Windows disks with sample sketches and scripts. Berkeley, CA. Key Curriculum Press.

VIDEOS

Apostol, Tom M. *The Story of Pi.* Defines, computes, and examines pi in terms of history and used in many fields. Includes exercises and projects. Reston, VA; NCTM.

NAME _______________ DATE _______

Parent Guide for Student Success

For use with Chapter 10

Chapter Overview One way that you can help your student succeed in Chapter 10 is by discussing the lesson goals in the chart below. When a lesson is completed, ask your student to interpret the lesson goals for you and to explain how the mathematics of the lesson relates to one of the key applications listed in the chart.

Lesson Title	*Lesson Goals*	*Key Applications*
10.1: Tangents to Circles	Identify segments and lines related to circles. Use properties of a tangent to a circle.	• Silo • Golf • Mexcaltitlán Island
10.2: Arcs and Chords	Use properties of arcs of circles and of chords of circles.	• Masonry Hammer • Time Zones • Avalanche Rescue Beacon
10.3: Inscribed Angles	Use inscribed angles to solve problems. Use properties of inscribed polygons.	• Theater Design • Carpenter's Square
10.4: Other Angle Relationships in Circles	Use angles formed by tangents and chords to solve problems in geometry. Use angles formed by lines that intersect a circle to solve problems.	• Views • Fireworks
10.5: Segment Lengths in Circles	Find the lengths of segments of chords. Find the lengths of segments of tangents and secants.	• Aquarium Tank • Designing a Logo • Global Positioning System
10.6: Equations of Circles	Write the equation of a circle. Use the equation of a circle and its graph to solve problems.	• Lighting • Cell Phones • Wankel Engine: Releaux Triangle
10.7: Locus	Draw the locus of points that satisfy a given condition. Draw the locus of points that satisfy two or more conditions.	• Locating an Epicenter • Dog Leash

Study Strategy

Answer Your Questions is the study strategy featured in Chapter 10 (see page 594). Encourage your student to make a list of questions whenever he or she doesn't understand something, especially problems assigned for homework. If you can't answer the questions, have your student ask the teacher and then record the answer in his or her math notebook.

NAME ______________________ DATE __________

Parent Guide for Student Success

For use with Chapter 10

Key Ideas **Your student can demonstrate understanding of key concepts by working through the following exercises with you.**

Lesson	*Exercise*
10.1	Point C is the center of a circle, point A is on the circle, and point B is outside the circle. $AC = 5$, $AB = 12$, and $BC = 14$. Is $\overleftrightarrow{AB}$ tangent to circle C? Explain.
10.2	Find $m\overset{\frown}{RS}$, $m\overset{\frown}{RPS}$, $m\overset{\frown}{PS}$, and $m\overset{\frown}{PR}$.
10.3	In the circle for the Exercise in Lesson 10.2, find $m\angle RPS$.
10.4	A meteorite flashes brightly 11 miles above sea level. Find the approximate measure of the arc that represents the part of Earth that can see the meteorite. Use 4000 miles for the radius of Earth.
10.5	You are standing about 4 feet from a circular silo. The distance from you to a point of tangency on the silo is 8 feet. Estimate the radius of the silo.
10.6	The listening area of a radio station is anywhere within 30 miles of the station's tower. If you are on a road 20 miles east and 20 miles south of the tower, can you hear the station? Explain.
10.7	You are given readings from three seismographs. At both $(2, -5)$ and $(1, 2)$ the epicenter is 5 miles away. At $(-2, 4)$ the epicenter is 6 miles away. Where is the epicenter?

Home Involvement Activity

You Will Need: A map of your town or neighborhood, compass, and straightedge.

Directions: Find a place you might live so three commuters in your family have the same distance to travel each day. On the map, mark the location of three places family members travel to the most. For example, mark your place of work and your student's school. If family members commute to only two places, mark the place where you shop most often. Use the properties of circles to construct the point that is equidistant from all three points. Is the central point in a residential area?

Answers

10.1: no; $\triangle ABC$ is not a right triangle because $5^2 + 12^2 \neq 14^2$. **10.2:** 80°; 280°; 140°; 140° **10.3:** 40° **10.4:** about 8° **10.5:** 6 ft **10.6:** yes; the point $(20, -20)$ is inside the listening area which is defined by the equation $x^2 + y^2 = 30^2$. **10.7:** at $(-2, -2)$

CHAPTER 10

NAME ______________________ DATE ________

Prerequisite Skills Review

For use before Chapter 10

EXAMPLE 1 Solving equations or systems of equations

Solve the equation or system of equations.

a. $3p^2 + 10 = 37$

b. $3x + 2y = 11$
$x + y = 2$

SOLUTION

a.
$$3p^2 + 10 = 37$$
$$3p^2 + 10 - 10 = 37 - 10$$
$$3p^2 = 27$$
$$p^2 = 9$$
$$p = \pm 3$$

b. $3x + 2y = 11$ $\quad 3x + 2y = 11$

$x + y = 2 \Rightarrow -2x - 2y = -4$ Multiply by -2.

$x = 7$ Add Equations.

$7 + y = 2$ Substitute 7 for x in the second equation.

$y = -5$ Solve for y.

Solution: $(7, -5)$

Exercises for Example 1

Solve the equation or system of equations.

1. $12^2 + x^2 = 13^2$

2. $x^2 + 4^2 = 5^2$

3. $(x + 3)^2 = x(x + 5)$

4. $2y^2 - 12 = 38$

5. $x - y = 2$
$4x + 3y = 36$

6. $2x + 3y = 4$
$4x + y = -2$

EXAMPLE 2 Solving Right Triangles

Solve the right triangle using the given information. Round decimals to the nearest tenth.

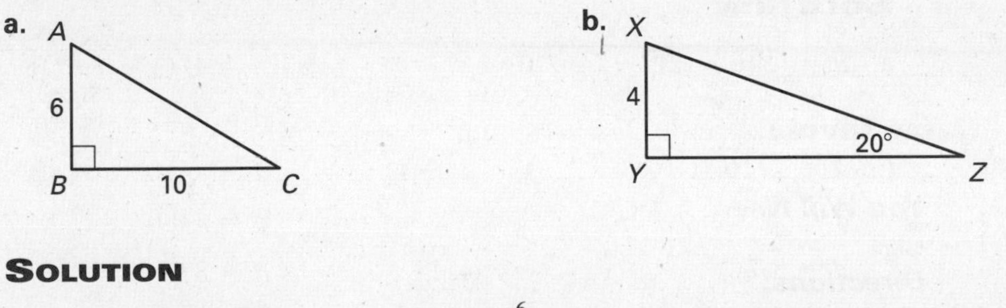

SOLUTION

a. $AC = \sqrt{6^2 + 10^2}$

$AC = \sqrt{36 + 100}$

$AC = \sqrt{136}$

$AC \approx 11.7$

$\tan C = \frac{6}{10}$

$m\angle C = \tan^{-1}\left(\frac{3}{5}\right)$

$m\angle C \approx 31°$

$m\angle A = 90° - 31°$

$m\angle A = 59°$

NAME _______________ DATE _______

Prerequisite Skills Review

For use after Chapter 10

b. $\tan 20° = \dfrac{4}{YZ}$

$YZ = \dfrac{4}{\tan 20°}$

$YZ = \dfrac{4}{0.36}$

$YZ = 11.0$

$XZ = \sqrt{4^2 + (11.0)^2}$

$XZ = \sqrt{16 + 121}$

$YZ = \sqrt{137}$

$YZ \approx 11.7$

$m\angle X = 90° - 20°$

$m\angle X = 70°$

Exercises for Example 2

Solve each right triangle. Round answers to the nearest tenth or nearest degree.

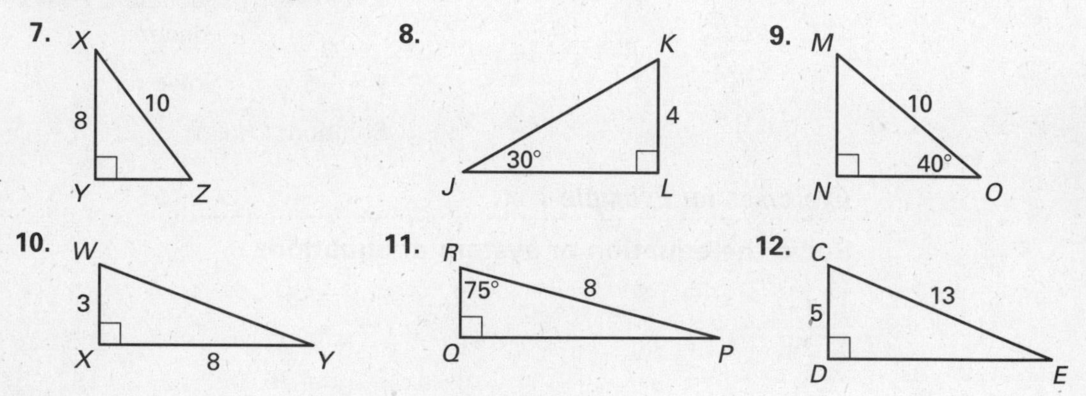

EXAMPLE 3 Finding the Length and Midpoint of a Segment

Find the length and the midpoint of $\overline{AB}$.

a. $A(2, 0), B(6, 4)$

b. $A(-1, 5), B(3, -3)$

SOLUTION

a. $d = \sqrt{(6 - 2)^2 + (4 - 0)^2}$

$d = \sqrt{4^2 + 4^2}$

$d = \sqrt{16 + 16}$

$d = \sqrt{32}$

$d \approx 5.7$

$m = \left(\dfrac{2 + 6}{2}, \dfrac{0 + 4}{2}\right)$

$m = (4, 2)$

b. $d = \sqrt{(3 - (-1))^2 + (-3 - 5)^2}$

$d = \sqrt{4^2 + (-8)^2}$

$d = \sqrt{16 + 64}$

$d = \sqrt{80}$

$d \approx 8.9$

$m = \left(\dfrac{-1 + 3}{2}, \dfrac{5 + (-3)}{2}\right)$

$m = (1, 1)$

Exercises for Example 3

Find the length and the midpoint of $\overline{AB}$. Write answers for length to the nearest tenth.

13. $A(4, 5), B(-2, 7)$

14. $A(0, 2), B(8, 0)$

15. $A(-6, 4), B(4, 2)$

16. $A(0, 5), B(12, 0)$

17. $A(-7, 3), B(4, 1)$

18. $A(5, 9), B(-3, -6)$

NAME ______________________ DATE __________

Strategies for Reading Mathematics

For use with Chapter 10

Strategy: Reading Postulates and Theorems

Postulates and theorems are an important part of geometry. You can find postulates and theorems throughout the book within chapters. There is also a list of postulates and a list of theorems at the back of the book. These lists allow you to see all of the postulates and theorems you have studied and proved. The postulates are listed on page 827, and the theorems on pages 828–832.

The *Arc Addition Postulate* is stated as shown below on page 827.

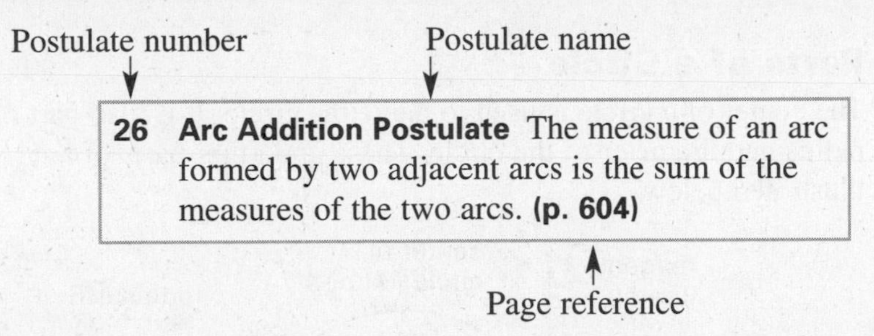

The theorems for Chapter 10 are listed on page 831. Theorem 10.1 is stated as shown below.

Theorem number

10.1 If a line is tangent to a circle, then it is perpendicular to the radius drawn to the point of tangency. **(p. 597)**

Page reference

STUDY TIP

Understanding Postulates and Theorems

Many of the postulates and theorems given in this book can be illustrated using diagrams. While the diagrams are not included in the back of the book, you can look on the page shown in parentheses to view a diagram associated with a postulate or theorem.

Questions

1. Why do you think a number is assigned to each postulate and theorem?
2. Why is the page reference listed in parentheses important?
3. Some postulates and theorems have names, while others do not. Why do you think this is so?
4. Why is it useful to have all of the postulates and theorems listed together?

NAME ______________________ DATE ________

Strategies for Reading Mathematics

For use with Chapter 10

Visual Glossary

The Study Guide on page 594 lists the vocabulary for Chapter 10 as well as review vocabulary from previous chapters. Use the page references on page 594 or the Glossary in the textbook to review key terms from prior chapters. Use the visual glossary below to help you understand some of the key vocabulary in Chapter 10. You may want to copy these diagrams into your notebook and refer to them as you complete the chapter.

Glossary

circle (p. 595) The set of all points in a plane that are equidistant from a given point, called the *center* of the circle.

chord (p. 595) A segment whose endpoints are on the circle.

secant (p. 595) A line that intersects a circle in two points.

tangent (p. 595) A line that intersects a circle in exactly one point, called the *point of tangency*.

central angle (p. 603) An angle whose vertex is the center of a circle.

major arc (p. 603) Part of a circle that measures between 180° and 360°.

minor arc (p. 603) Part of a circle that measures less than 180°.

inscribed angle (p. 613) An angle whose vertex is on a circle and whose sides contain chords of the circle.

intercepted arc (p. 613) The arc that lies in the interior of an inscribed angle and has endpoints on the angle.

Parts of a Circle

The center of a circle is used to name the circle. It is also part of each radius and diameter of the circle. These and other parts of a circle are illustrated below.

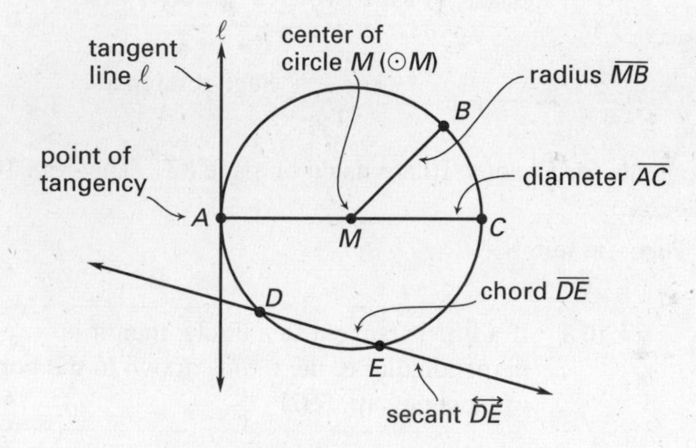

Angles and Arcs of a Circle

An arc is part of a circle associated with a central angle of the circle. Minor arcs are named by their endpoints. Major arcs are named by their endpoints and by a point on the arc.

An inscribed angle has its vertex on the circle, while the vertex of a central angle is the center of the circle. If an angle is inscribed in a circle, then it intercepts an arc.

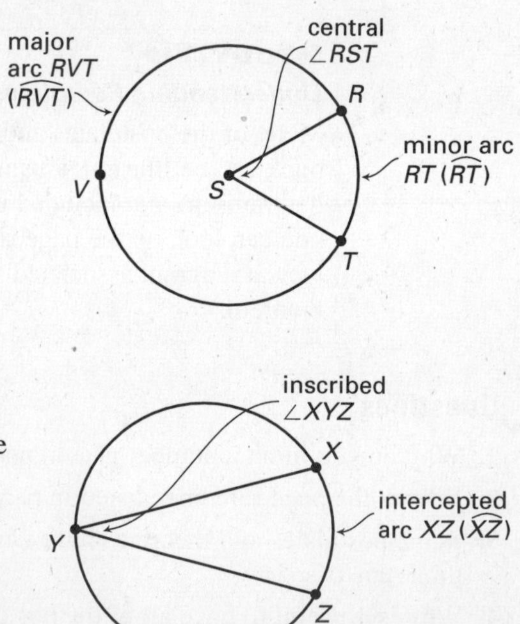

LESSON
10.1

TEACHER'S NAME ____________________ CLASS ______ ROOM ______ DATE ______

Lesson Plan

2-day lesson (See *Pacing the Chapter,* TE pages 592C–592D) **For use with pages 595–602**

1. **Identify segments and lines related to circles.**
2. **Use properties of a tangent to a circle.**

State/Local Objectives ____

____

✓ Check the items you wish to use for this lesson.

STARTING OPTIONS

____ Prerequisite Skills Review: CRB pages 5–6
____ Strategies for Reading Mathematics: CRB pages 7–8
____ Homework Check: TE page 576: Answer Transparencies
____ Warm-Up or Daily Homework Quiz: TE pages 595 and 580, CRB page 11, or Transparencies

TEACHING OPTIONS

____ Lesson Opener (Application): CRB page 12 or Transparencies
____ Technology Activity with Keystrokes: CRB pages 13–15
____ Examples: Day 1: 1–3, SE pages 595–596; Day 2: 4–7, SE pages 597–598
____ Extra Examples: Day 1: TE page 596 or Transp.; Day 2: TE pages 597–598 or Transp.; Internet
____ Closure Question: TE page 598
____ Guided Practice: SE page 599 Day 1: Exs. 1–2, 4; Day 2: Exs. 3, 5–8

APPLY/HOMEWORK

Homework Assignment

____ Basic Day 1: 9–35; Day 2: 36–49, 54–56, 58–70 even
____ Average Day 1: 9–35; Day 2: 36–49, 54–56, 58–70 even
____ Advanced Day 1: 9–35; Day 2: 36–57, 58–70 even

Reteaching the Lesson

____ Practice Masters: CRB pages 16–18 (Level A, Level B, Level C)
____ Reteaching with Practice: CRB pages 19–20 or Practice Workbook with Examples
____ Personal Student Tutor

Extending the Lesson

____ Applications (Real-Life): CRB page 22
____ Challenge: SE page 602; CRB page 23 or Internet

ASSESSMENT OPTIONS

____ Checkpoint Exercises: Day 1: TE pages 596–597 or Transp.; Day 2: TE pages 597–598 or Transp.
____ Daily Homework Quiz (10.1): TE page 602, CRB page 26, or Transparencies
____ Standardized Test Practice: SE page 602; TE page 602; STP Workbook; Transparencies

Notes ____

____

____

LESSON 10.1

TEACHER'S NAME ____________________ CLASS ______ ROOM ______ DATE ______

Lesson Plan for Block Scheduling

1-day lesson (See *Pacing the Chapter,* TE pages 592C–592D)

For use with pages 595–602

GOALS
1. Identify segments and lines related to circles.
2. Use properties of a tangent to a circle.

State/Local Objectives ____________________

CHAPTER PACING GUIDE	
Day	Lesson
1	**10.1 (all)**
2	10.2 (all)
3	10.3 (all)
4	10.4 (all); 10.5 (begin)
5	10.5 (end); 10.6 (all)
6	10.7 (all)
7	Review Ch. 10; Assess Ch. 10

✓ Check the items you wish to use for this lesson.

STARTING OPTIONS

____ Prerequisite Skills Review: CRB pages 5–6
____ Strategies for Reading Mathematics: CRB pages 7–8
____ Homework Check: TE page 576: Answer Transparencies
____ Warm-Up or Daily Homework Quiz: TE pages 595 and 580, CRB page 11, or Transparencies

TEACHING OPTIONS

____ Lesson Opener (Application): CRB page 12 or Transparencies
____ Technology Activity with Keystrokes: CRB pages 13–15
____ Examples 1–7: SE pages 595–598
____ Extra Examples: TE pages 596–598 or Transparencies; Internet
____ Closure Question: TE page 598
____ Guided Practice Exercises: SE page 599

APPLY/HOMEWORK

Homework Assignment

____ Block Schedule: 9–49, 54–56, 58–70 even

Reteaching the Lesson

____ Practice Masters: CRB pages 16–18 (Level A, Level B, Level C)
____ Reteaching with Practice: CRB pages 19–20 or Practice Workbook with Examples
____ Personal Student Tutor

Extending the Lesson

____ Applications (Real-Life): CRB page 22
____ Challenge: SE page 602; CRB page 23 or Internet

ASSESSMENT OPTIONS

____ Checkpoint Exercises: TE pages 596–598 or Transparencies
____ Daily Homework Quiz (10.1): TE page 602, CRB page 26, or Transparencies
____ Standardized Test Practice: SE page 602; TE page 602; STP Workbook; Transparencies

Notes ____________________

LESSON 10.1

NAME ______________________________ DATE __________

Available as a transparency

WARM-UP EXERCISES

For use before Lesson 10.1, pages 595–602

Solve the equation.

1. $2x = x + 5$

2. $3w + 4 = 5w - 8$

3. $h^2 + 4 = 40$

4. $m^2 - 16 = 32$

5. $(y + 2)^2 + 4 = 29$

Lesson 10.1

DAILY HOMEWORK QUIZ

For use after Lesson 9.7, pages 573–580

Draw vector $\overrightarrow{PQ}$ in a coordinate plane. Write the component form of the vector and find its magnitude. Round your answer to the nearest tenth.

1. $P(-2, -3), Q(1, 4)$

2. $P(-4, 5,) \; Q(3, -1)$

Let $\vec{a} = \langle 4, -2 \rangle$, $\vec{b} = \langle -3, 1 \rangle$, and $\vec{c} = \langle 7, 2 \rangle$. Find the given sum.

3. $\vec{b} + \vec{c}$

4. $\vec{a} + \vec{c}$

NAME ______________________ DATE __________

Available as a transparency

Application Lesson Opener

For use with pages 595–602

The steam engine was the first effective machine for producing power independently of human and animal muscle, wind, or flowing water. James Watt was a Scottish inventor and engineer whose improvements to the steam engine in the late 1700s were key technological foundations of the Industrial Revolution. One of his improvements was a "double-acting" engine, in which steam powered a piston in both directions of its movement. This made it possible to attach a rod to a crank to produce rotary (circular) motion. This, in turn, made the engine useful for running machinery or even driving the wheels of a carriage or the paddles of a riverboat. By the end of the 18th century, the steam engines designed by Watt were providing power for factories, mills, and pumps both in Europe and America.

The diagram shows some key parts of a simple steam engine. A sliding valve (not shown) lets steam in on the left (pushing the piston to the right) and then lets steam in on the right (pushing the piston to the left.)

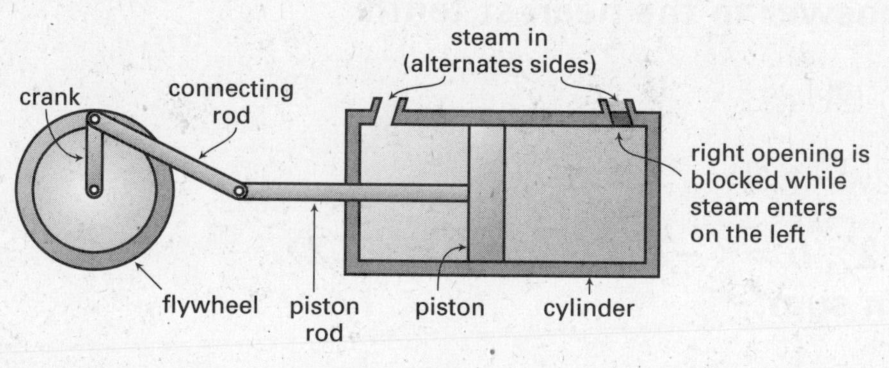

1. As the piston moves back and forth in the cylinder, the flywheel rotates. In the diagram, the crank points up while the piston is in the center of the cylinder. Draw diagrams of 90°, 180°, and 270° clockwise rotations of this crank. In each diagram label the position of the piston. (Note that the piston rod remains horizontal.)

2. The length of the crank (or the radius of the flywheel) is called the *throw* of the crank. How does the throw of the crank relate to the distance the piston moves? Give an example.

NAME ______________________________ DATE __________

Technology Activity

For use with pages 595–602

GOAL Verifying statements about tangent lines to circles using geometry software

Geometry software can be used to verify statements about tangent lines to circles. For example, you could use geometry software to construct the diagram described. Then, you could use the software's tools to verify the statement about the segment length.

Given: l is tangent to $\odot Q$.

Prove: $l \perp \overline{QP}$

Activity

1. Draw $\odot Q$ centered at the origin with a radius of 1 unit.
2. Draw radius $\overline{QP}$ where P is at $(1, 0)$.
3. Draw vertical line l tangent to $\odot Q$ at point P.
4. Measure $\angle P$. Is $l \perp \overline{QP}$?

Exercise

1. Use geometry software to verify the following.

   Given: $\overleftrightarrow{SR}$ and $\overleftrightarrow{ST}$ are tangent to $\odot P$.

   Prove: $\overline{SR} \cong \overline{ST}$

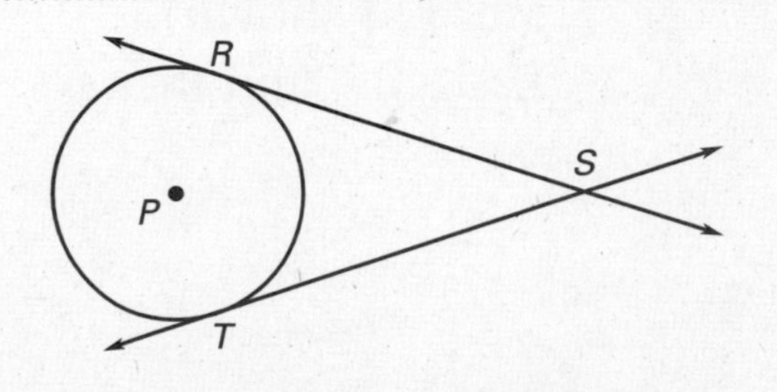

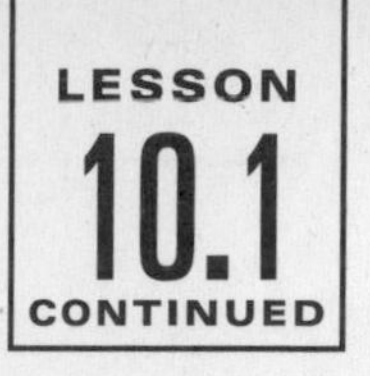

NAME ______________________ DATE __________

Technology Activity Keystrokes

For use with pages 595–602

TI-92

1. Turn on the axes and the grid.

   F8 9 (Set Coordinate Axes to RECTANGULAR and Grid to ON.) ENTER

   Draw ⊙Q.

   F3 1 (Move cursor to the origin.) ENTER (Move cursor to (1, 0).) ENTER

2. Draw radius $\overline{QP}$.

   F2 5 (Move cursor to center of ⊙Q.) ENTER (Move cursor to (1, 0).) ENTER

3. Draw line l tangent to ⊙Q at point P.

   F2 4 (Move cursor to P.) ENTER (Move cursor to (1, 1).) ENTER

4. Measure $\angle P$.

   F6 3 (Move cursor to line l.) ENTER (Move cursor to angle P.) ENTER (Move cursor to segment $\overline{QP}$.) ENTER

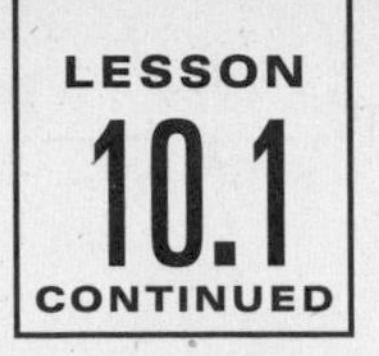

NAME ______________________ DATE ____________

Technology Activity Keystrokes

For use with pages 595–602

SKETCHPAD

1. Turn on the axes and the grid. Choose **Snap To Grid** from the **Graph** menu. Draw $\odot Q$ using the compass tool, placing the center of the circle at the origin and drawing the circle so that the radius is 1 unit.
2. Draw radius $\overline{QP}$, where P is at $(1, 0)$, using the segment straightedge tool.
3. Draw vertical line l tangent to $\odot Q$ at point P. Use the line straightedge tool and draw the line through point P and $(1, 1)$.
4. Measure $\angle P$. Use the selection arrow tool to select a point on line l, hold down the shift key, and select P and the center of $\odot Q$. Then choose **Angle** from the **Measure** menu.

LESSON 10.1

NAME ____________________ DATE ________

Practice A

For use with pages 595–602

The diameter of a circle is given. Find the radius.

1. $d = 6$ in.
2. $d = 24$ cm
3. $d = 15$ ft
4. $d = 9$ in.

The radius of a circle is given. Find the diameter.

5. $r = 11$ cm
6. $r = 8$ ft
7. $r = 10$ in.
8. $r = 4.6$ cm

Match the notation with the term that best describes it.

9. D A. Center
10. $\overleftrightarrow{FH}$ B. Chord
11. $\overline{CD}$ C. Diameter
12. $\overline{AB}$ D. Radius
13. C E. Point of tangency
14. $\overline{AD}$ F. Common external tangent
15. $\overleftrightarrow{AB}$ G. Common internal tangent
16. $\overleftrightarrow{DE}$ H. Secant

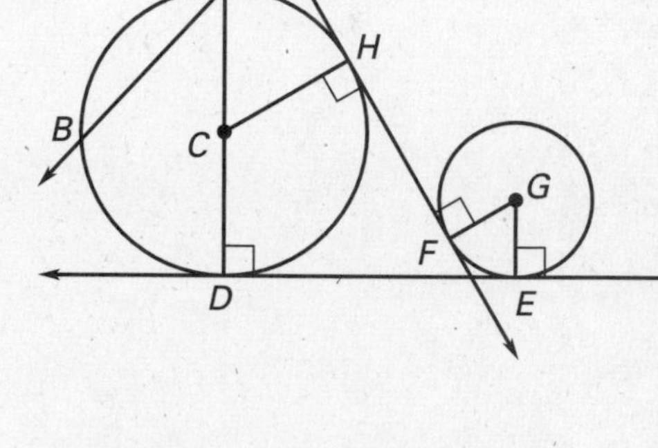

Use the diagram at the right.

17. What are the center and radius of $\odot A$?
18. What are the center and radius of $\odot B$?
19. Describe the intersection of the two circles.
20. Describe all the common tangents of the two circles.
21. Are the two circles congruent? Explain.

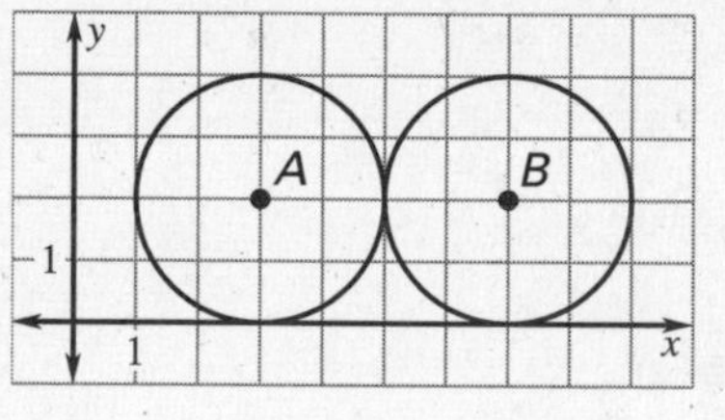

Tell whether $\overleftrightarrow{AB}$ is tangent to $\odot C$. Explain your reasoning.

22. 23.

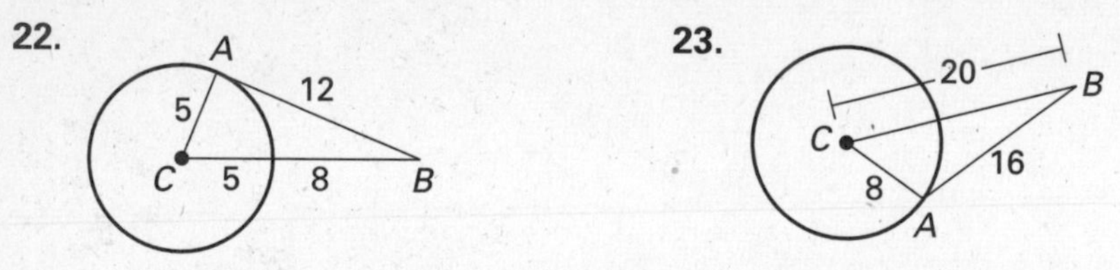

24. ***Baseball Stadium*** The shape of the outfield fence in a baseball stadium is that of a quarter circle. If the distance from home plate to the wall is 330 feet, what is the radius of the entire circle? What is the diameter of the circle?

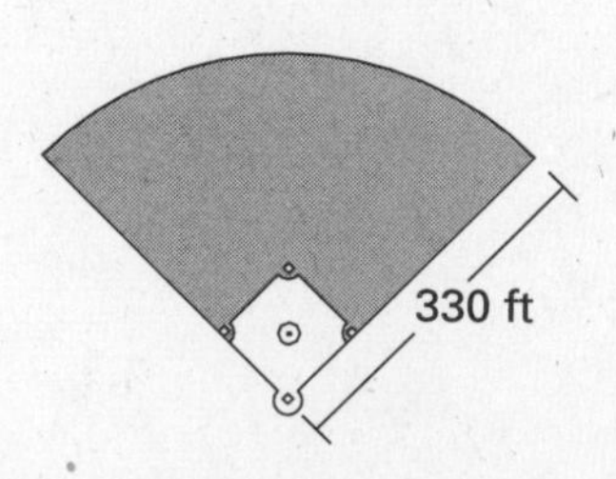

LESSON 10.1

NAME ______________________________ DATE ____________

Reteaching with Practice

For use with pages 595–602

GOAL **Identify segments and lines related to circles and use properties of a tangent to a circle**

Vocabulary

A **circle** is the set of all points in a plane that are equidistant from a given point, called the **center** of the circle.

The distance from the center to a point on the circle is the **radius** of the circle.

Two circles are **congruent** if they have the same radius.

The distance across the circle, through its center, is the **diameter** of the circle.

A **radius** is a segment whose endpoints are the center of the circle and a point on the circle.

A **chord** is a segment whose endpoints are points on the circle.

A **diameter** is a chord that passes through the center of the circle.

A **secant** is a line that intersects a circle in two points.

A **tangent** is a line in the plane of a circle that intersects the circle in exactly one point.

Theorem 10.1
If a line is tangent to a circle, then it is perpendicular to the radius drawn to the point of tangency.

Theorem 10.2
In a plane, if a line is perpendicular to a radius of a circle at its endpoint on the circle, then the line is tangent to the circle.

Theorem 10.3
If two segments from the same exterior point are tangent to a circle, then they are congruent.

EXAMPLE 1 ***Identifying Special Segments and Lines***

Tell whether the line or segment is best described as a *chord,* a *secant,* a *tangent,* a *diameter,* or a *radius* of $\odot C$.

a. $\overline{HC}$ **b.** $\overleftrightarrow{DG}$

c. $\overline{BE}$ **d.** $\overleftrightarrow{AF}$

e. $\overline{BH}$

H C G A B E F D

Solution

a. $\overline{HC}$ is a radius because one of its endpoints, H, is a point on the circle and the other endpoint, C, is the circle's center.

b. $\overleftrightarrow{DG}$ is a tangent because it intersects the circle in one point.

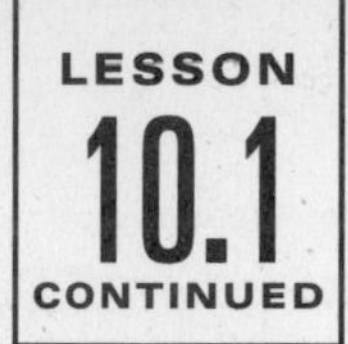

NAME ______________________________ DATE ____________

Reteaching with Practice

For use with pages 595–602

c. $\overline{BE}$ is a chord because its endpoints are on the circle.

d. $\overleftrightarrow{AF}$ is a secant because it intersects the circle in two points.

e. $\overline{BH}$ is a diameter because its endpoints are on the circle and it contains the center C.

Exercises for Example 1

In Exercises 1–8, tell whether the line or segment is best described as a *chord*, a *secant*, a *tangent*, a *diameter*, or a *radius* of $\odot C$.

1. $\overline{AB}$ **2.** $\overleftrightarrow{DE}$

3. $\overline{DC}$ **4.** $\overline{DE}$

5. $\overleftrightarrow{FG}$ **6.** $\overline{CG}$

7. $\overline{EG}$ **8.** $\overline{EC}$

EXAMPLE 2 Using Properties of Tangents

$\overleftrightarrow{AB}$ and $\overleftrightarrow{AD}$ are tangent to $\odot C$. Find the value of x.

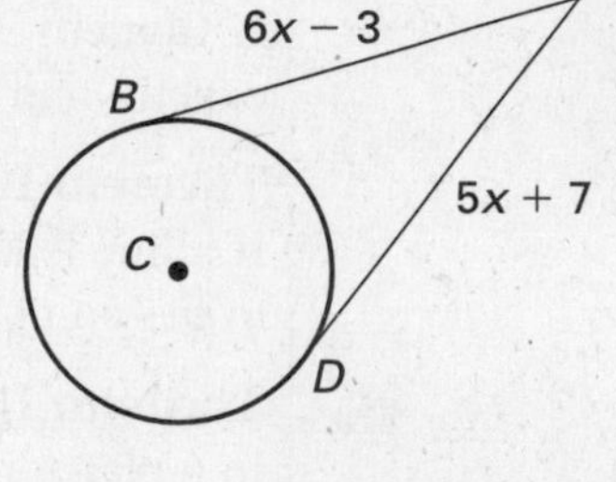

SOLUTION

$AB = AD$	Two tangent segments from the same point are congruent.
$6x - 3 = 5x + 7$	Substitute.
$6x = 5x + 10$	Add 3 to each side.
$x = 10$	Subtract $5x$ from each side.

Exercises for Example 2

In Exercises 9–11, $\overleftrightarrow{AB}$ and $\overleftrightarrow{AD}$ are tangent to $\odot C$. Find the value of x.

9.

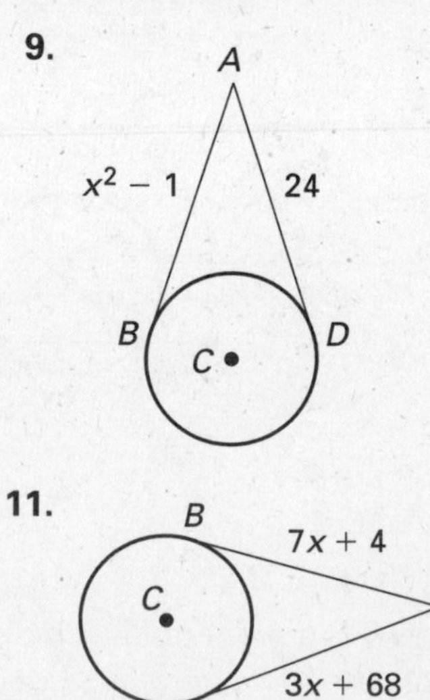

10.

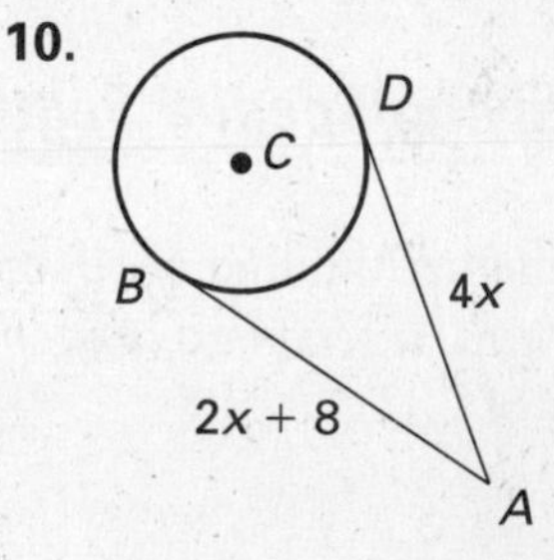

11.

LESSON 10.1

NAME ______________________________________ DATE __________

Quick Catch-Up for Absent Students

For use with pages 595–602

The items checked below were covered in class on (date missed) ____________

Lesson 10.1: Tangents to Circles

____ **Goal 1:** Identify segments and lines related to circles. (pp. 595–596)

Material Covered:

____ Example 1: Identifying Special Segments and Lines

____ Example 2: Identifying Common Tangents

____ Example 3: Circles in Coordinate Geometry

Vocabulary:

circle, p. 595
center, p. 595
radius, p. 595
congruent, p. 595
diameter, p. 595
chord, p. 595
secant, p. 595
tangent, p. 595
tangent circles, p. 596
concentric, p. 596
common tangent, p. 596
interior of a circle, p. 596
exterior of a circle, p. 596

____ **Goal 2:** Use properties of a tangent to a circle. (pp. 597–598)

Material Covered:

____ Student Help: Study Tip

____ Example 4: Verifying a Tangent to a Circle

____ Student Help: Skills Review

____ Example 5: Finding the Radius of a Circle

____ Example 6: Proof of Theorem 10.3

____ Example 7: Using Properties of Tangents

Vocabulary:

point of tangency, p. 597

____ Other (specify) ____

____

Homework and Additional Learning Support

____ Textbook (specify) pp. 599–602

____

____ Internet: Extra Examples at www.mcdougallittell.com

____ *Reteaching with Practice* worksheet (specify exercises) ______________

____ *Personal Student Tutor* for Lesson 10.1

NAME ______________________________ DATE __________

Real-Life Application: When Will I Ever Use This?

For use with pages 595–602

Lesson 10.1

Satellite Transmissions

The field of telecommunications is increasingly relying on satellites. First a tool of the military, they are utilized now by news services, telephone companies, the Internet, and cable and satellite television. Once practically a monopoly by the United States, satellite launches are now done by other countries such as Australia, who compete with NASA for the lucrative business of launching commercial satellites for telecommunication companies.

Satellites have only a limited range because of the curvature of Earth. The transmissions can only reach the points that are tangent to Earth from the satellite. Since it is important to create point to point communications, a network of satellites must be launched to cover a desired area. Some satellites are fixed, or stationary in the sky. These satellites are orbiting Earth at a speed that allows them to hover over the same location. This allows more reliable communication because it is always available to a certain area of coverage. Some orbit at a speed that allows them to travel around the globe and appear at many locations several times a day. These offer more coverage, but are available to locations only for certain parts of the day.

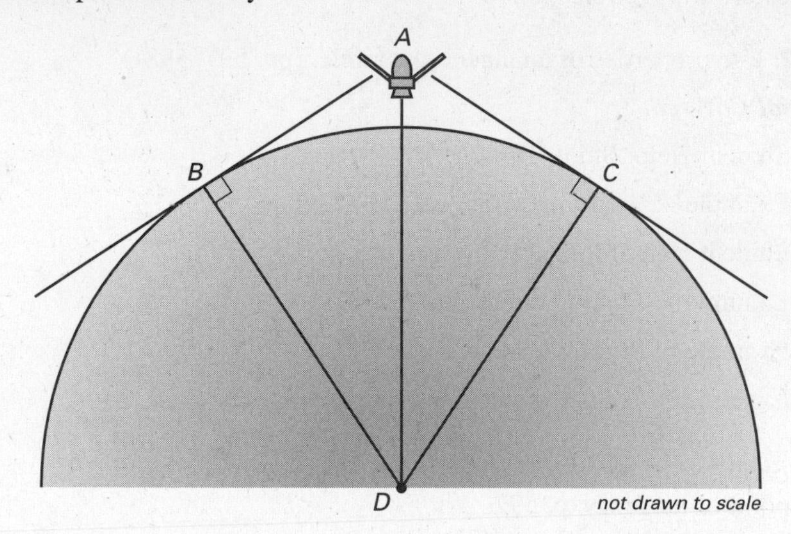

In Exercises 1–5, use the diagram above where the satellite (point *A*) is stationary.

1. What are the points of tangency from the satellite?
2. Identify the radii.
3. What theorem says that angles *ABD* and *ACD* are right angles?
4. Prove triangles *ADC* and *ADB* are congruent.
5. What theorem says that $\overline{AB}$ must be congruent to $\overline{AC}$?

LESSON 10.1

NAME ______________________ DATE __________

Challenge: Skills and Applications

For use with pages 595–602

1. In the diagram, C is the center of both circles, and the radii of the circles are 8 and 17. If $\overline{JL}$ is tangent to the circle of radius 8, find the length of $\overline{JL}$.

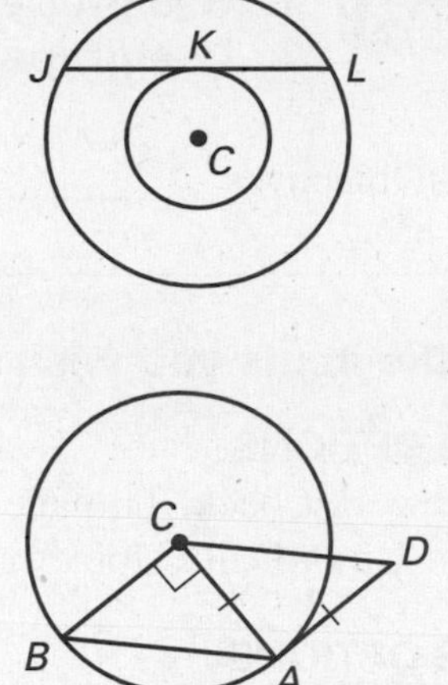

2. Refer to the diagram. Write a two-column proof.

Given: C is the center of the circle, $\overline{AC} \cong \overline{AD}$, $\overline{AC} \perp \overline{BC}$, and $\overline{AD}$ is tangent to the circle at A.

Prove: $ABCD$ is a parallelogram.

3. If $XY = 18$, $YZ = 14$, and $XZ = 20$, find the radius of each circle.

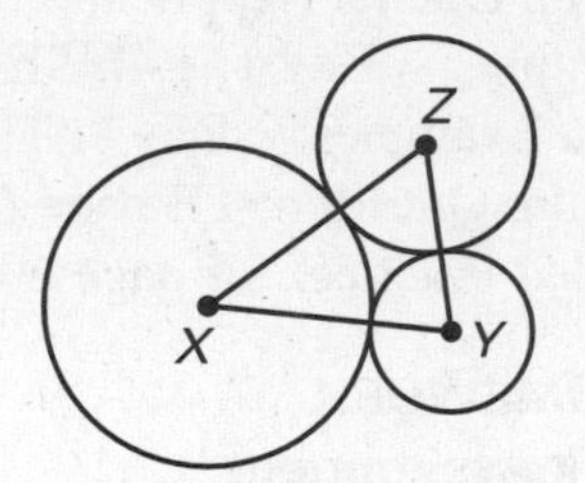

4. In the diagram, $\overline{IK}$ and $\overline{JL}$ are tangent to the circles at I and L, respectively. If $\overline{IJ}$ and $\overline{KL}$ are radii, $IJ = 15$, and $IK = 20$, find JL.

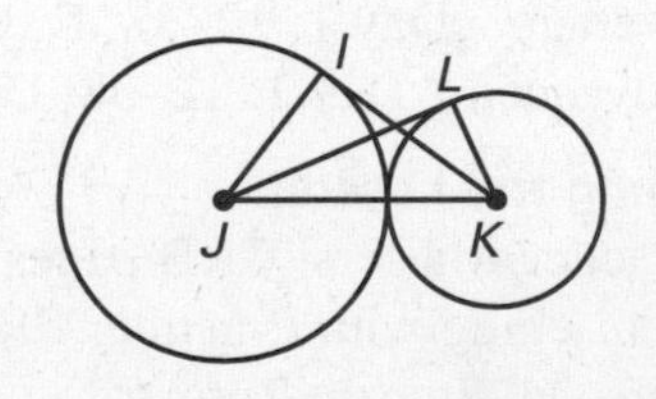

5. The diagram shows two pulleys with a belt wrapped snugly around the pulleys so that one pulley can drive the other. $\overline{RS}$ is tangent to the circles at R and S, respectively, $\overline{QT}$ is perpendicular to $\overline{SP}$, and Q and P are centers of the circles. Let $QR = 2$ in., $PS = 8$ in., and $PQ = 12$ in.

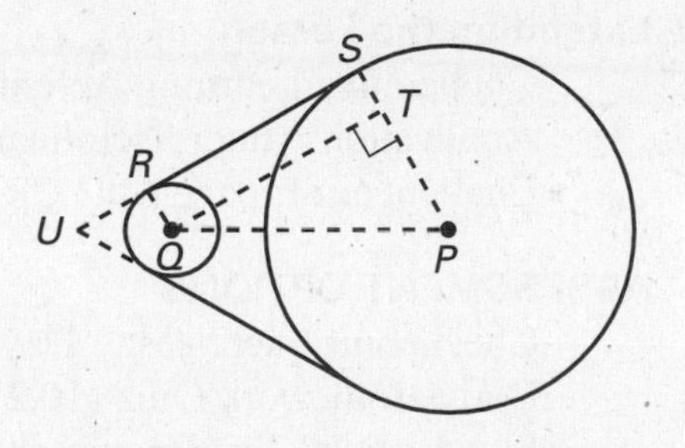

a. Write a paragraph proof to show that $QRST$ is a rectangle.

b. Find RS.

c. Find $m\angle P$.

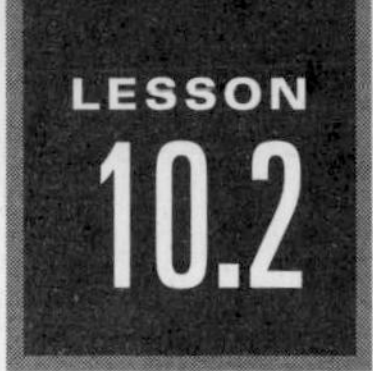

TEACHER'S NAME ______________________ CLASS ______ ROOM ______ DATE ______

Lesson Plan

2-day lesson (See *Pacing the Chapter,* TE pages 592C–592D) **For use with pages 603–611**

GOALS
1. **Use properties of arcs of circles.**
2. **Use properties of chords of circles.**

State/Local Objectives ____

____

✓ Check the items you wish to use for this lesson.

STARTING OPTIONS

____ Homework Check: TE page 599: Answer Transparencies
____ Warm-Up or Daily Homework Quiz: TE pages 603 and 602, CRB page 26, or Transparencies

TEACHING OPTIONS

____ Motivating the Lesson: TE page 604
____ Lesson Opener (Application): CRB page 27 or Transparencies
____ Examples: Day 1: 1–4, SE pages 603–605; Day 2: 5–7, SE pages 605–606
____ Extra Examples: Day 1: TE pages 604–605 or Transp.; Day 2: TE pages 605–606 or Transp.
____ Closure Question: TE page 606
____ Guided Practice: SE page 607 Day 1: Exs. 1–8; Day 2: Exs. 9–11

APPLY/HOMEWORK

Homework Assignment

____ Basic Day 1: 12–38; Day 2: 39–48, 54–57, 63–68, 70–77
____ Average Day 1: 12–38; Day 2: 39–51, 53–57, 59, 63–68, 70–77
____ Advanced Day 1: 12–38; Day 2: 39–51, 53–57, 59, 63–77

Reteaching the Lesson

____ Practice Masters: CRB pages 28–30 (Level A, Level B, Level C)
____ Reteaching with Practice: CRB pages 31–32 or Practice Workbook with Examples
____ Personal Student Tutor

Extending the Lesson

____ Cooperative Learning Activity: CRB page 34
____ Applications (Interdisciplinary): CRB page 35
____ Challenge: SE page 611; CRB page 36 or Internet

ASSESSMENT OPTIONS

____ Checkpoint Exercises: Day 1: TE page 605 or Transp.; Day 2: TE pages 605–606 or Transp.
____ Daily Homework Quiz (10.2): TE page 611, CRB page 39, or Transparencies
____ Standardized Test Practice: SE page 611; TE page 611; STP Workbook; Transparencies

Notes ____

____

____

LESSON 10.2

TEACHER'S NAME ______________ CLASS ______ ROOM ______ DATE ______

Lesson Plan for Block Scheduling

1-day lesson (See *Pacing the Chapter,* TE pages 592C–592D) For use with pages 603–611

GOALS

1. Use properties of arcs of circles.
2. Use properties of chords of circles.

State/Local Objectives ______________

CHAPTER PACING GUIDE	
Day	Lesson
1	10.1 (all)
2	**10.2 (all)**
3	10.3 (all)
4	10.4 (all); 10.5 (begin)
5	10.5 (end); 10.6 (all)
6	10.7 (all)
7	Review Ch. 10; Assess Ch. 10

✓ Check the items you wish to use for this lesson.

STARTING OPTIONS

____ Homework Check: TE page 599: Answer Transparencies
____ Warm-Up or Daily Homework Quiz: TE pages 603 and 602, CRB page 26, or Transparencies

TEACHING OPTIONS

____ Motivating the Lesson: TE page 604
____ Lesson Opener (Application): CRB page 27 or Transparencies
____ Examples 1–7: SE pages 603–606
____ Extra Examples: TE pages 604–606 or Transparencies
____ Closure Question: TE page 606
____ Guided Practice Exercises: SE page 607

APPLY/HOMEWORK

Homework Assignment

____ Block Schedule: 12–51, 53–57, 59, 63–68, 70–77

Reteaching the Lesson

____ Practice Masters: CRB pages 28–30 (Level A, Level B, Level C)
____ Reteaching with Practice: CRB pages 31–32 or Practice Workbook with Examples
____ Personal Student Tutor

Extending the Lesson

____ Cooperative Learning Activity: CRB page 34
____ Applications (Interdisciplinary): CRB page 35
____ Challenge: SE page 611; CRB page 36 or Internet

ASSESSMENT OPTIONS

____ Checkpoint Exercises: TE pages 605–606 or Transparencies
____ Daily Homework Quiz (10.2): TE page 611, CRB page 39, or Transparencies
____ Standardized Test Practice: SE page 611; TE page 611; STP Workbook; Transparencies

Notes ______________

LESSON 10.2

NAME ______________________ DATE __________

Available as a transparency

WARM-UP EXERCISES

For use before Lesson 10.2, pages 603–611

Solve the equation.

1. $3x = x + 50$

2. $y + 5y + 66 = 360$

3. $x + 14x = 180$

4. $a^2 + 16 = 25$

5. $3w + 4w + 5w = 360$

DAILY HOMEWORK QUIZ

For use after Lesson 10.1, pages 595–602

The radius of ⊙*C* is given. Find the diameter of ⊙*C*.

1. 13 ft

2. 3.2 in.

Tell whether $\overline{AB}$ is tangent to ⊙*C*.

3.

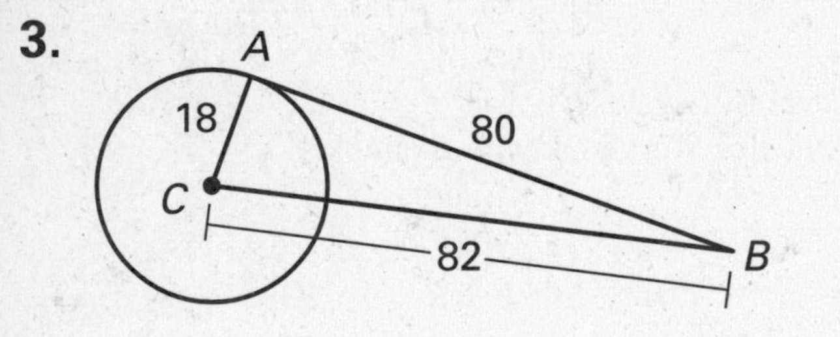

$\overline{AB}$ and $\overline{AD}$ are tangent to ⊙*C*. Find the value of *x*.

4.

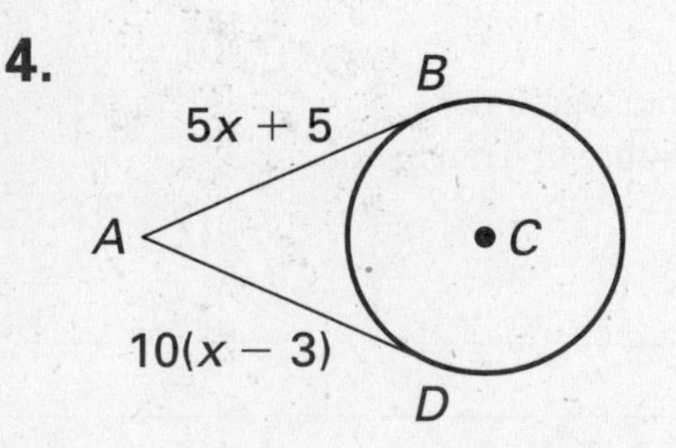

Lesson 10.2

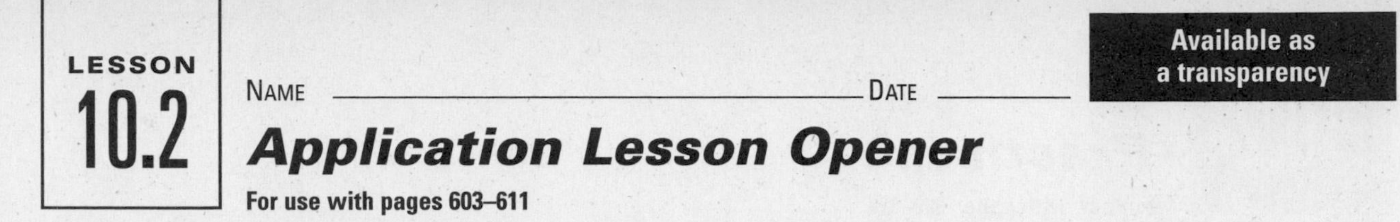

LESSON **10.2**

NAME _______________ DATE _______

Application Lesson Opener

For use with pages 603–611

Available as a transparency

You will need: • graph paper • ruler • compass • protractor

An engineer designs a curve in a road that is an arc of a circle as follows: Given two points A and B, the engineer locates a point C on the perpendicular bisector of $\overline{AB}$, then draws a circle with center C and radius AC. The portion of the circle from A to B is an arc that represents the curved road between A and B.

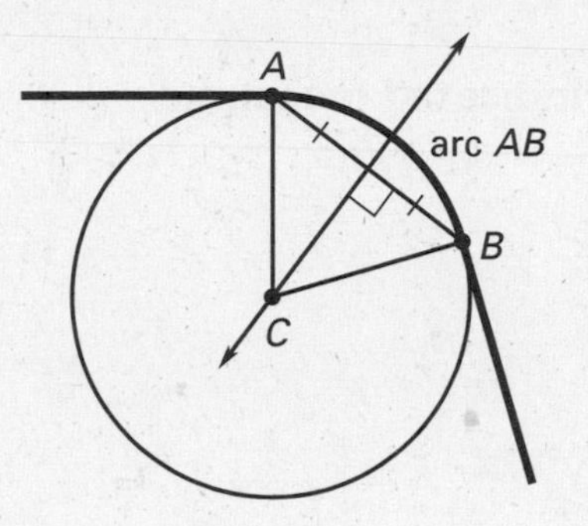

1. Why does the engineer look for a point on the perpendicular bisector of $\overline{AB}$?

2. Graph the points $A(2, 2)$ and $B(14, 6)$ on a coordinate plane. Use the engineer's method to design a curved road between A and B. (You will need a compass.) Label the coordinates of point C. Find the length of the radius of $\odot C$. Find the measure of $\angle ACB$. How many choices are there for the location of C? Explain.

3. On the same graph, label M as the midpoint of $\overline{AB}$. Draw a circle with center M and radius AM. Do you think this circle gives a suitable arc for a curved road between A and B? Explain.

4. Draw several other arcs between A and B. As the radius of the circle drawn gets longer, what happens to the length of the arc between A and B? What happens to the measure of $\angle ACB$?

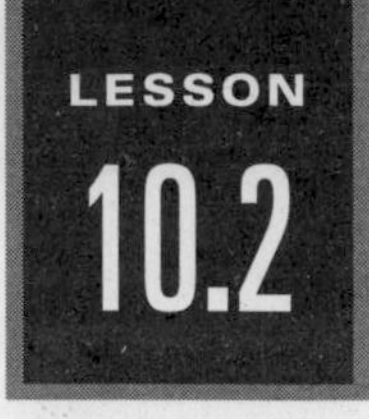

NAME ______________________________ DATE __________

Practice A

For use with pages 603–611

Determine whether the arc is a *minor arc*, a *major arc*, or a *semicircle* of ⊙*C*.

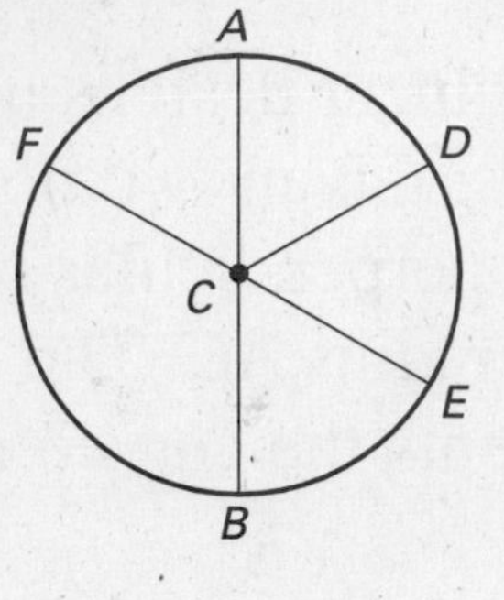

1. $\overset{\frown}{AE}$
2. $\overset{\frown}{AEB}$
3. $\overset{\frown}{FDE}$
4. $\overset{\frown}{DFB}$
5. $\overset{\frown}{FA}$
6. $\overset{\frown}{BE}$
7. $\overset{\frown}{BDA}$
8. $\overset{\frown}{FB}$

$\overline{MQ}$ and $\overline{NR}$ are diameters. Find the indicated measure.

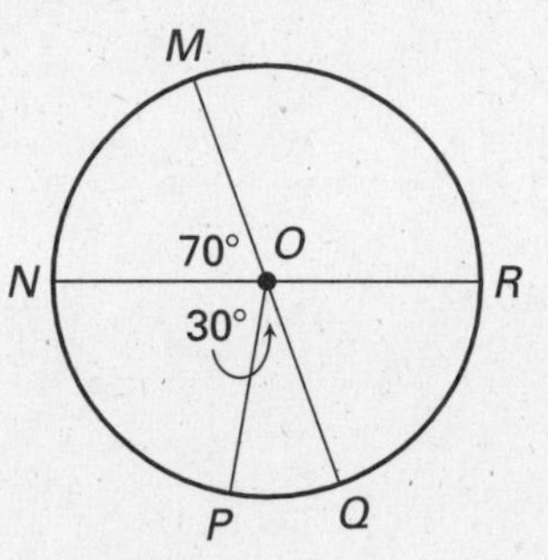

9. $m\overset{\frown}{MN}$
10. $m\overset{\frown}{NQ}$
11. $m\overset{\frown}{NQR}$
12. $m\overset{\frown}{MRP}$
13. $m\overset{\frown}{QR}$
14. $m\overset{\frown}{MR}$
15. $m\overset{\frown}{QMR}$
16. $m\overset{\frown}{PQ}$
17. $m\overset{\frown}{PRN}$
18. $m\overset{\frown}{MQN}$

Find the measure of $\overset{\frown}{MN}$.

19.

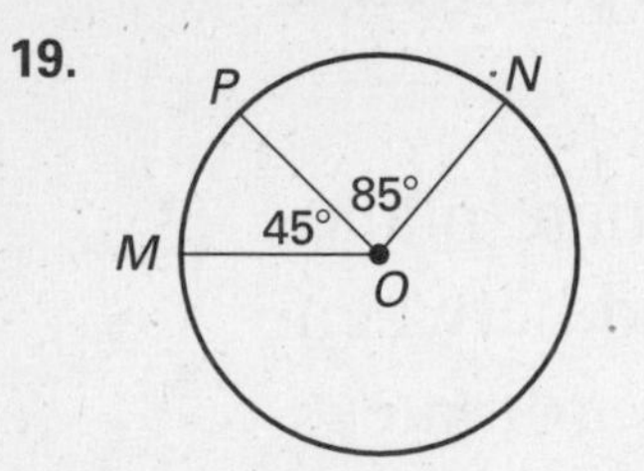

20.

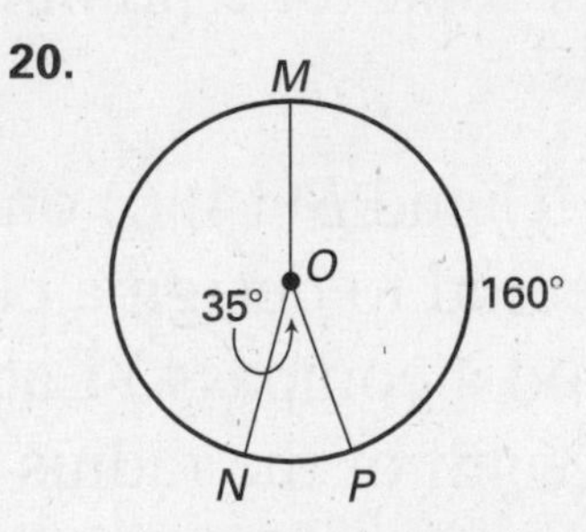

21.

Q
100°
P
140°
O
N
M

What can you conclude about the diagram? State a postulate or theorem that justifies your answer.

22.

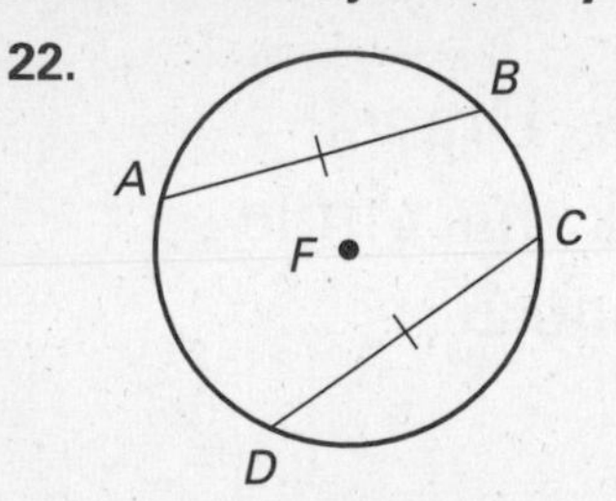

23.

A
C
F
140°
140°
B

24.

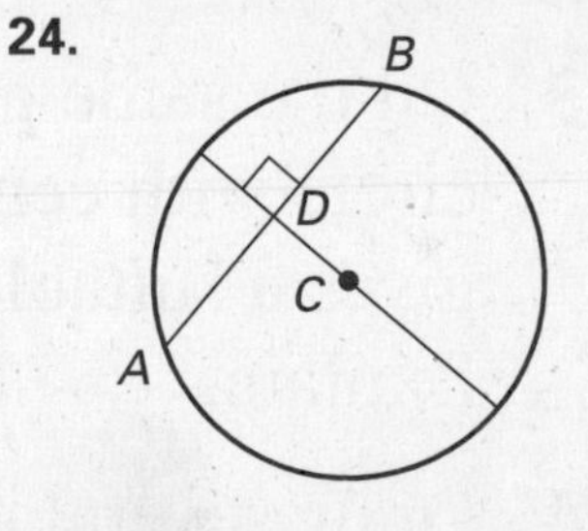

Find the indicated measure for ⊙*P*.

25. $DC =$ ___?___

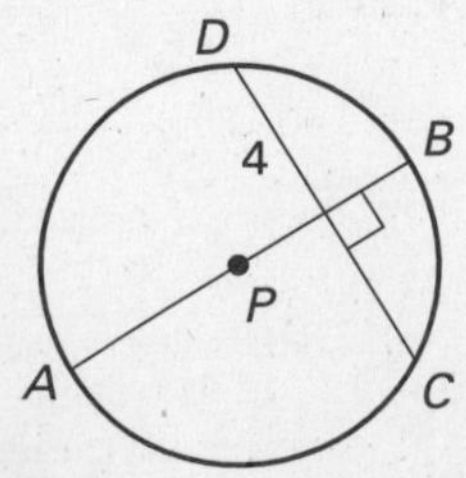

26. $AD =$ ___?___

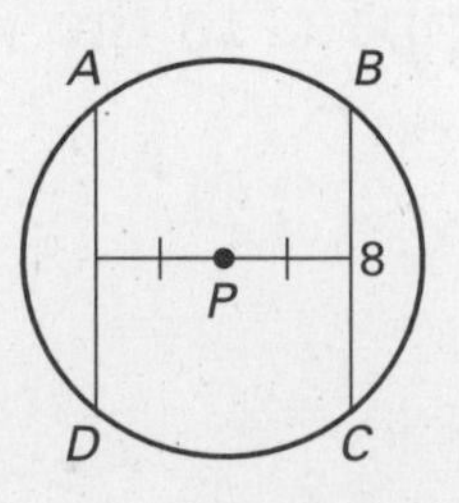

27. $EC =$ ___?___

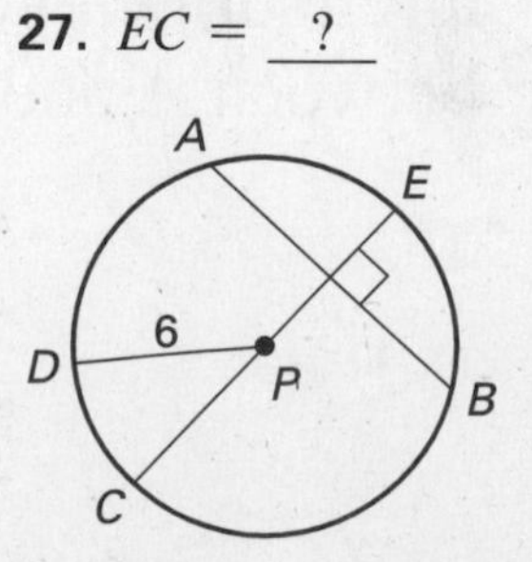

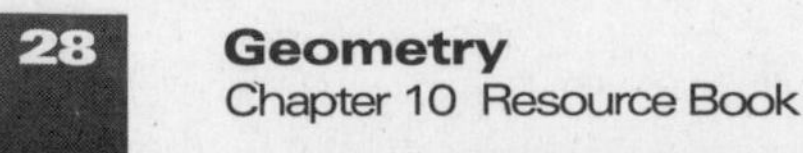

NAME ______________________ DATE ________

Practice B

For use with pages 603–611

Determine whether the arc is a *minor arc*, a *major arc*, or a *semicircle* of $\odot C$.

1. $\overset{\frown}{AE}$
2. $\overset{\frown}{ADB}$
3. $\overset{\frown}{FDE}$
4. $\overset{\frown}{DFB}$
5. $\overset{\frown}{FA}$
6. $\overset{\frown}{BE}$
7. $\overset{\frown}{BDA}$
8. $\overset{\frown}{FB}$

$\overline{MQ}$ and $\overline{NR}$ are diameters. Find the indicated measure.

9. $m\overset{\frown}{MN}$
10. $m\overset{\frown}{NQ}$
11. $m\overset{\frown}{NQR}$
12. $m\overset{\frown}{MRP}$
13. $m\overset{\frown}{QR}$
14. $m\overset{\frown}{MR}$
15. $m\overset{\frown}{QMR}$
16. $m\overset{\frown}{PQ}$
17. $m\overset{\frown}{PRN}$
18. $m\overset{\frown}{MQN}$

Find the measure of $\overset{\frown}{MN}$.

19.

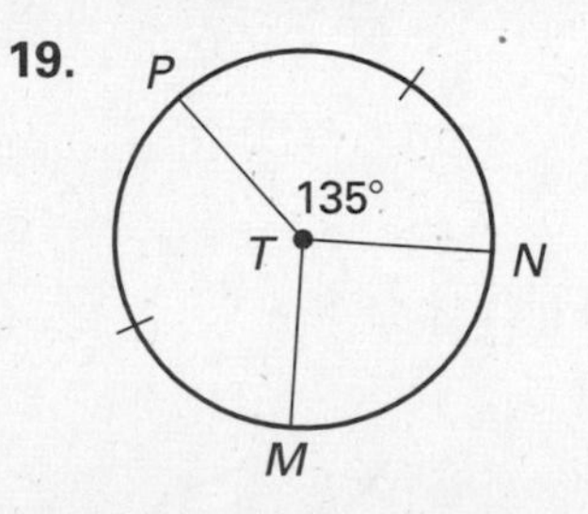

20.

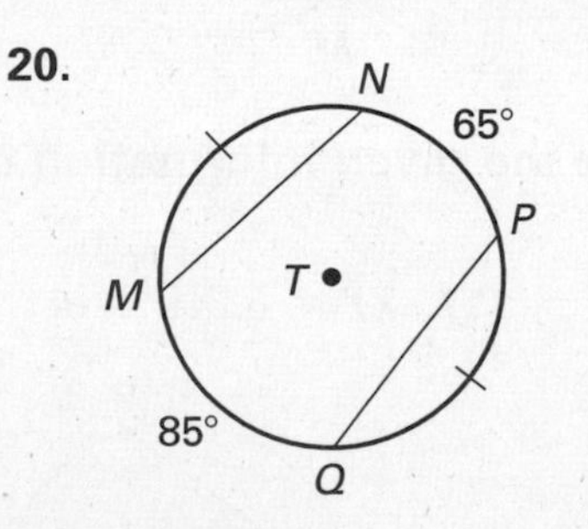

21.

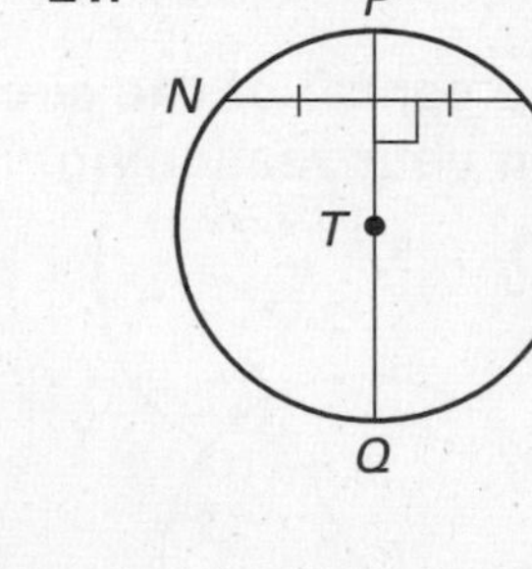

Use the figures to match the chord or arc with a congruent chord or arc.

22. $\overset{\frown}{FB}$
23. $\overline{AF}$
24. $\overset{\frown}{BC}$
25. $\overline{EC}$
26. $\overset{\frown}{DC}$
27. $\overline{PD}$

A. $\overset{\frown}{FE}$
B. $\overset{\frown}{ED}$
C. $\overset{\frown}{EC}$
D. $\overline{AB}$
E. $\overline{BF}$
F. $\overline{PA}$

Find the indicated measure for $\odot P$.

28. $FC = \underline{\quad ? \quad}$

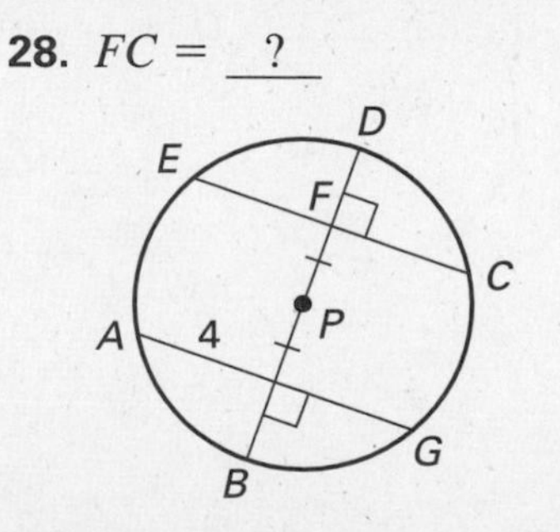

29. $m\overset{\frown}{BC} = 50°$, $\overset{\frown}{AB} \cong \overset{\frown}{ED}$, $m\overset{\frown}{AE} = \underline{\quad ? \quad}$

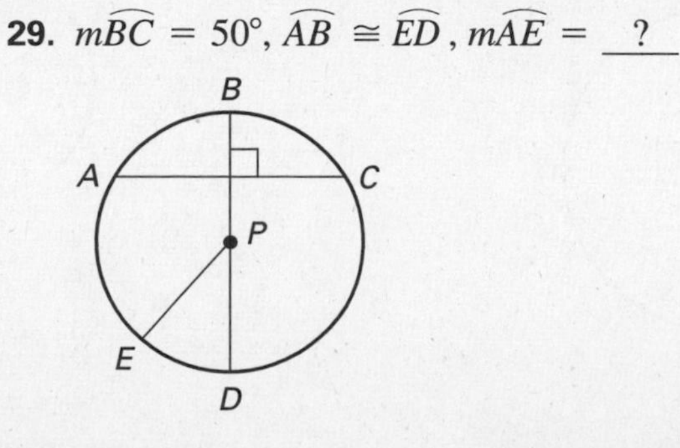

NAME ______________________________ DATE __________

Practice C

For use with pages 603–611

$\overline{MQ}$ and $\overline{NR}$ are diameters. Find the indicated measures.

1. $m\overset{\frown}{MN}$
2. $m\overset{\frown}{NQ}$
3. $m\overset{\frown}{NQR}$
4. $m\overset{\frown}{MRP}$
5. $m\overset{\frown}{PN}$
6. $m\overset{\frown}{MNQ}$
7. $m\overset{\frown}{QR}$
8. $m\overset{\frown}{MR}$
9. $m\overset{\frown}{QMR}$
10. $m\overset{\frown}{PQ}$
11. $m\overset{\frown}{PRN}$
12. $m\overset{\frown}{MQN}$

Find the measure of $\overset{\frown}{MN}$.

13.

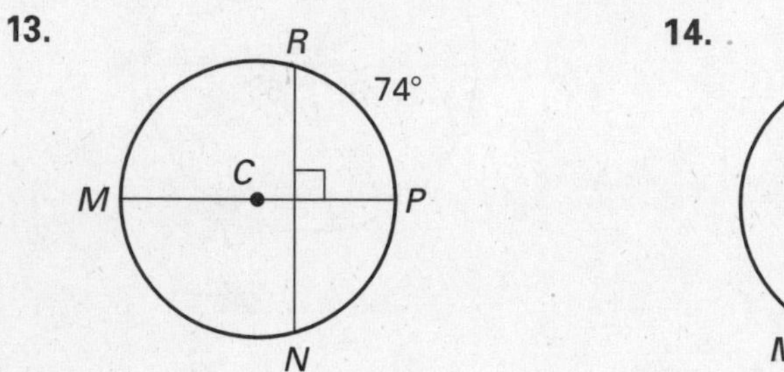

14.

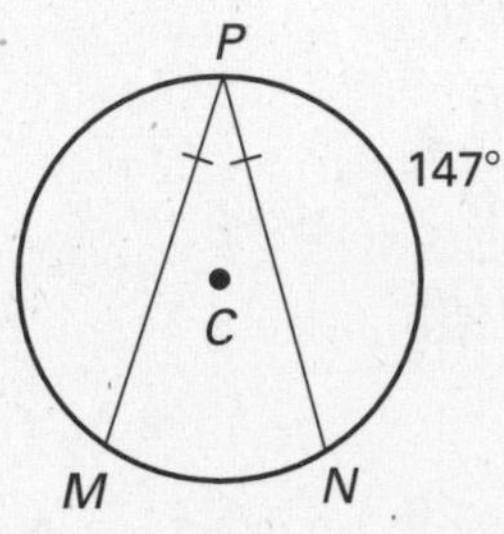

15.

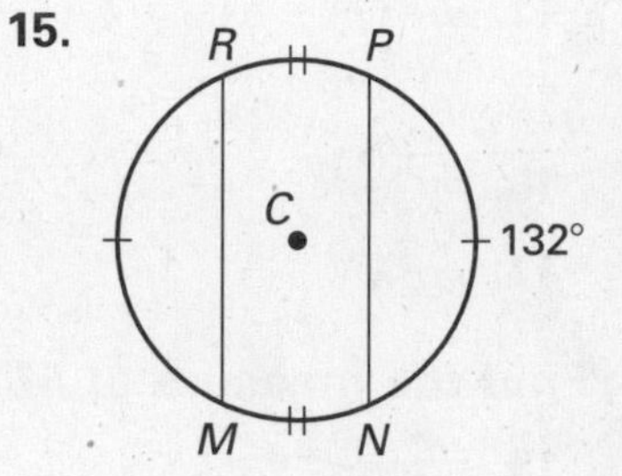

***P* is the center of the circle. Use the given information to find *XY*. Explain your reasoning.**

16. $ZY = 3$

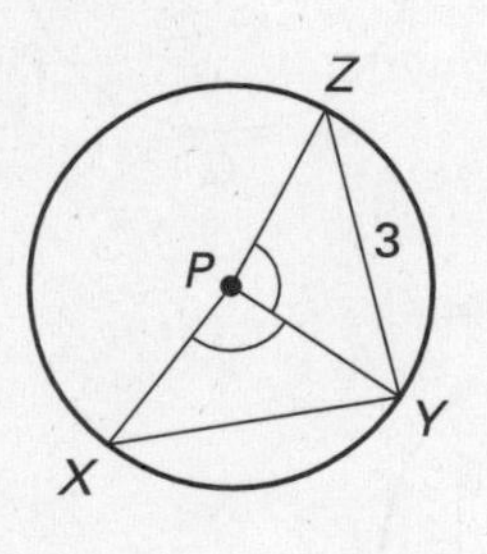

17. $ZY = 6, XW = 4$

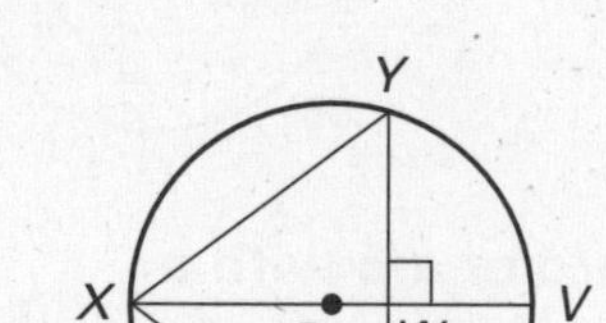

18. $CA = 3$

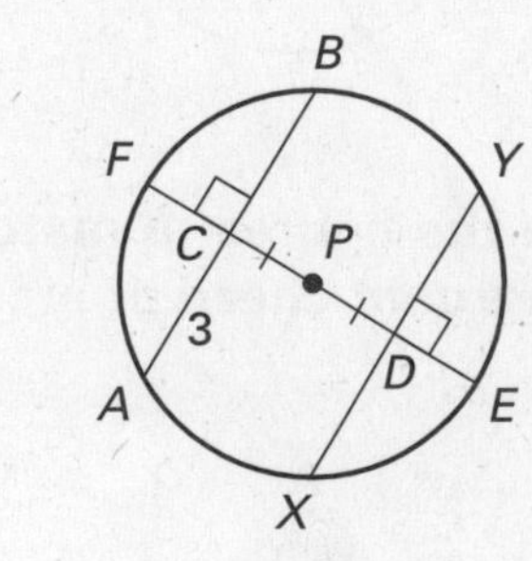

Write a two-column proof or a paragraph proof.

19. **Given:** $\odot P$, $\overline{AG} \perp \overline{CD}$

$\overline{AG}$ is a diameter of $\odot P$.

Prove: $\overline{AC} \cong \overline{AD}$

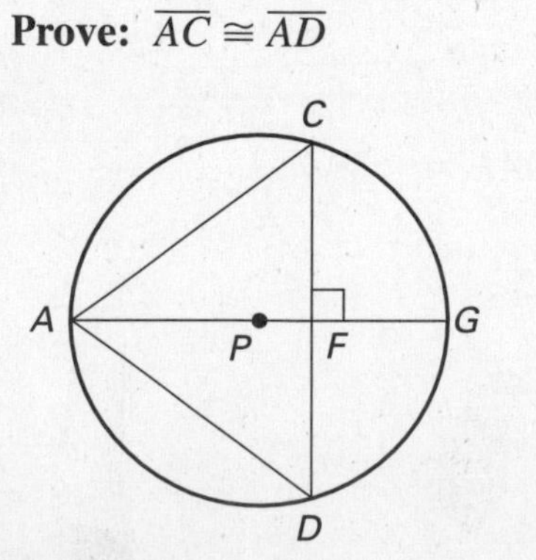

20. **Given:** $\odot P$, $\odot Q$, $\overline{CQ} \cong \overline{AP}$

$\overset{\frown}{AB} \cong \overset{\frown}{CD}$

Prove: $\triangle APB \cong \triangle CQD$

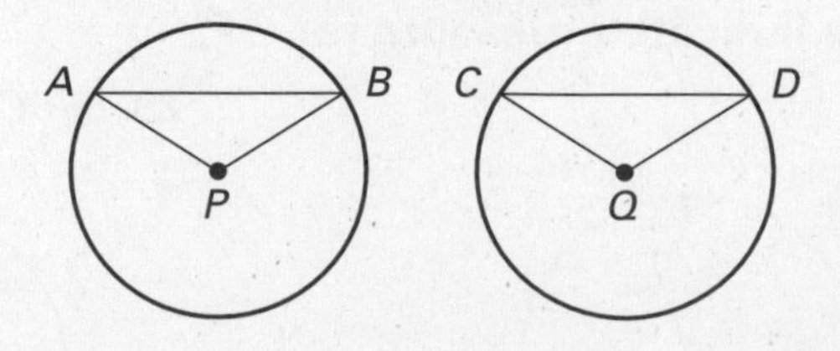

NAME ______________________________ DATE ____________

Reteaching with Practice

For use with pages 603–611

GOAL Use properties of arcs of circles and use properties of chords of circles

Vocabulary

In a plane, an angle whose vertex is the center of a circle is a **central angle** of the circle.

If the measure of a central angle, $\angle APB$, is less than 180°, then A and B and the points of $\odot P$ in the interior of $\angle APB$ form a **minor arc** of the circle.

The points A and B and the points of $\odot P$ in the *exterior* of $\angle APB$ form a **major arc** of the circle. If the endpoints of an arc are the endpoints of a diameter, then the arc is a **semicircle.**

The **measure of a minor arc** is defined to be the measure of its central angle.

The **measure of a major arc** is defined as the difference between 360° and the measure of its associated minor arc.

Two arcs of the same circle or of congruent circles are **congruent arcs** if they have the same measure.

Postulate 26 Arc Addition Postulate
The measure of an arc formed by two adjacent arcs is the sum of the measures of the two arcs.

Theorem 10.4
In the same circle, or in congruent circles, two minor arcs are congruent if and only if their corresponding chords are congruent.

Theorem 10.5
If a diameter of a circle is perpendicular to a chord, then the diameter bisects the chord and its arc.

Theorem 10.6
If one chord is a perpendicular bisector of another chord, then the first chord is a diameter.

Theorem 10.7
In the same circle, or in congruent circles, two chords are congruent if and only if they are equidistant from the center.

EXAMPLE *Finding Measures of Arcs*

Find the measure of each arc of $\odot C$.

a. $\overset{\frown}{AD}$ **b.** $\overset{\frown}{ADB}$

c. $\overset{\frown}{DBA}$ **d.** $\overset{\frown}{BD}$

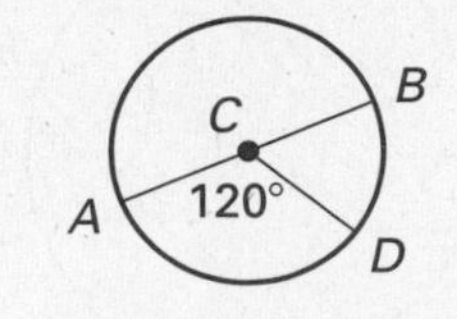

Solution

a. $\overset{\frown}{AD}$ is a minor arc, so $m\overset{\frown}{AD} = m\angle ACD = 120°$.

b. $\overset{\frown}{ADB}$ is a semicircle, so $m\overset{\frown}{ADB} = 180°$.

NAME ______________________ DATE __________

Reteaching with Practice

For use with pages 603–611

c. $\overset{\frown}{DBA}$ is a major arc, so $m\overset{\frown}{DBA} = 360° - 120° = 240°$.

d. $\overset{\frown}{BD}$ is a minor arc, so $m\overset{\frown}{BD} = m\angle BCD$. Because $\angle BCD$ and $\angle ACD$ form a linear pair, $m\angle BCD = 180° - m\angle ACD = 180° - 120° = 60°$. So $m\overset{\frown}{BD} = 60°$.

Exercises for Example 1

Find the measure of the arcs of the given circle.

1. Find the measure of each arc of $\odot C$.

a. $\overset{\frown}{ADB}$ **b.** $\overset{\frown}{AD}$

c. $\overset{\frown}{DB}$ **d.** $\overset{\frown}{DBA}$

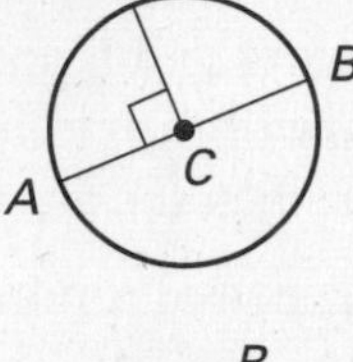

2. Find the measure of each arc of $\odot Q$.

a. $\overset{\frown}{PR}$ **b.** $\overset{\frown}{PRS}$

c. $\overset{\frown}{PS}$ **d.** $\overset{\frown}{RSP}$

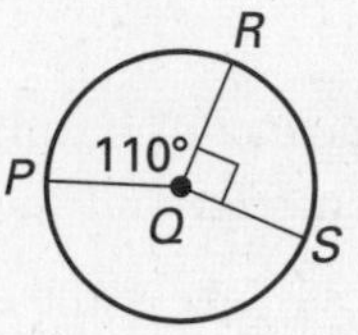

EXAMPLE 2 Using Theorem 10.7

$PS = 12$, $TV = 12$, and $SQ = 7$. Find QU.

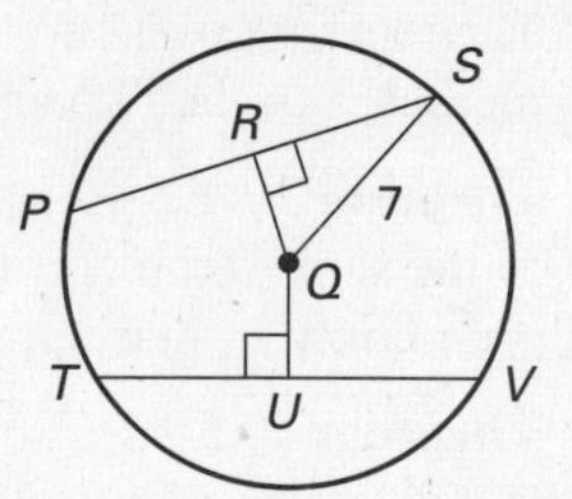

SOLUTION

Because $\overline{PS} \cong \overline{TV}$, they are equidistant from the center by Theorem 10.7. To find QU, first find QR. $\overline{QR} \perp \overline{PS}$, so $\overline{QR}$ bisects $\overline{PS}$. Because $PS = 12$, $RS = \frac{12}{2} = 6$. Now look at $\triangle QRS$ which is a right triangle. Use the Pythagorean Theorem to find QR. $QR = \sqrt{QS^2 - RS^2} = \sqrt{7^2 - 6^2} = \sqrt{13}$. Because $\overline{QR} \cong \overline{QU}$, $QR = QU = \sqrt{13}$.

Exercises for Example 2

Use the given information to find the value of *x*.

3. $AB = DE = 10$, radius $= 6$

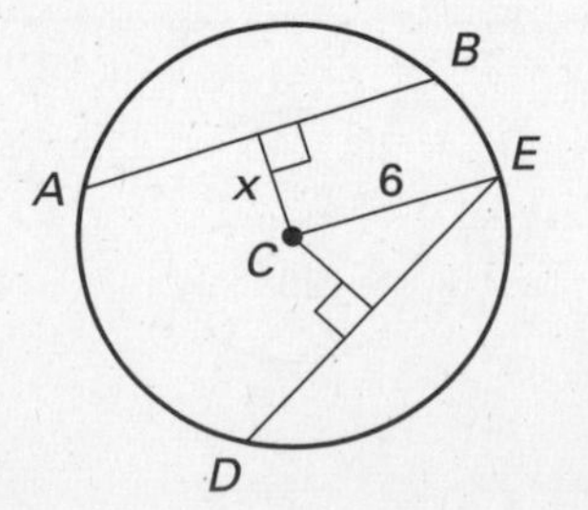

4. $QV = 2$, $QU = 2$, $SU = 3$

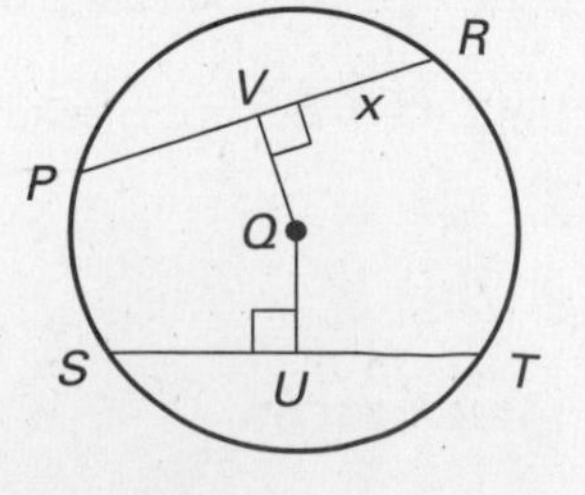

LESSON 10.2

NAME ______________________________ DATE __________

Quick Catch-Up for Absent Students

For use with pages 603–611

The items checked below were covered in class on (date missed) __________

Lesson 10.2: Arcs and Chords

____ **Goal 1:** Use properties of arcs of circles. (pp. 603–604)

Material Covered:

____ Example 1: Finding Measures of Arcs

____ Example 2: Finding Measures of Arcs

____ Example 3: Identifying Congruent Arcs

Vocabulary:

central angle, p. 603
minor arc, p. 603
major arc, p. 603
semicircle, p. 603
measure of a minor arc, p. 603
measure of a major arc, p. 603
congruent arcs, p. 604

____ **Goal 2:** Use properties of chords of circles. (pp. 605–606)

Material Covered:

____ Example 4: Using Theorem 10.4

____ Example 5: Finding the Center of a Circle

____ Example 6: Using Properties of Chords

____ Student Help: Look Back

____ Example 7: Using Theorem 10.7

____ Other (specify) ______________________________

____

Homework and Additional Learning Support

____ Textbook (specify) pp. 607–611

____

____ *Reteaching with Practice* worksheet (specify exercises) __________

____ *Personal Student Tutor* for Lesson 10.2

Lesson 10.2

LESSON

10.2

NAME _______________ DATE _______

Cooperative Learning Activity

For use with pages 603–611

GOAL **To investigate the properties of chords and arcs of a circle and to understand how these properties relate to each other and to central angles**

Materials: compass, ruler, protractor, paper, pencil

Exploring Arcs and Chords

The use of arcs and chords is instrumental in the understanding of circles. Knowing the measure of these two parts of a circle allows for the discovery of other important information about the circle.

Instructions

1. Find a circular object in the classroom that can be used as a guide to draw a circle. Each member of the group should draw a different sized circle (if no round objects are found, use a compass to draw the circles.) Label the center of the circle O.
2. Construct two congruent chords in the circle (use the compass to insure the chords are congruent). Label the chords $\overline{JK}$ and $\overline{LM}$.
3. Construct radii $\overline{OJ}$, $\overline{OK}$, $\overline{OL}$, and $\overline{OM}$.
4. Measure $\angle KOJ$ and $\angle LOM$ with your protractor.
5. Construct a second circle. Mark the center of the circle.
6. Construct two nonparallel congruent chords that are not diameters.
7. Construct the perpendiculars from the center to each chord.
8. With the compass, compare the distances from the center of the chords (measure along the perpendicular from the center to the chord).

Analyzing the Results

1. How do the measures of $\angle KOJ$ and $\angle LOM$ compare?
2. How do the distances from the center to the chords compare?
3. Does the size of the circle have any effect on the answers to Exercises 1 and 2?

LESSON 10.2

NAME ______________________ DATE ________

Interdisciplinary Application

For use with pages 603–611

Wheelwrights

HISTORY The United States expanded in territory greatly during the 1800s. As pioneers opened new parts of the west and northwest territories, there followed thousands of settlers. Many traveled along the famous Oregon Trail which started in "jumping off" places like Independence, Missouri, and stretched 2000 miles to Oregon and California. Starting about 1841, the most popular vehicle chosen for the journey was the covered wagon.

Wheelwrights were men who specialized in the manufacture and repair of wheels used for buggies and covered wagons. In the Old West there did not exist the standard sizes for wheels that we have today. Each wheelwright made wagon wheels according to his own specifications. Making a new wheel was challenging because the new wheel would have to be the same size as the old one. Wagon wheels did not go flat. They broke. Often the wheelwright would have to reconstruct the size of a wheel from broken remains. When the remnant was too small to directly measure the radius, a method could be used utilizing Theorem 10.6. The partial wheel was laid on the dirt and chords to the circle were made. The perpendicular bisectors of these chords were made and the center of the circle found. The following exercises will lead you through this method.

In Exercises 1–4, use the diagram below of the partial wagon wheel.

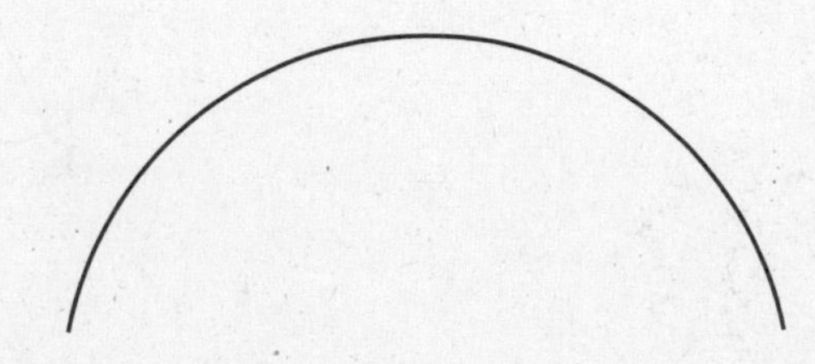

1. In the circle above, draw two chords $\frac{1}{2}$ inch long and mark the middle of these chords. Using a protractor, draw perpendicular bisectors through the midpoints that intersect inside the circle.
2. Each of the perpendicular bisectors run along which chord in the circle?
3. The bisectors intersect each other at what point in the circle? How do you know this?
4. How can you complete the circle? Complete the circle.

Lesson 10.2

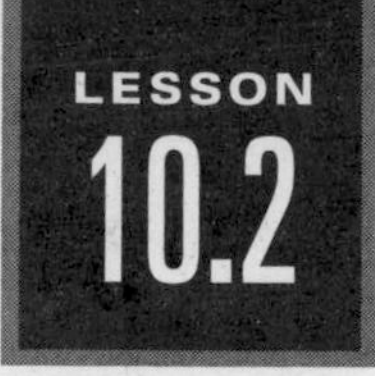

NAME ______________________ DATE __________

Challenge: Skills and Applications

For use with pages 603–611

1. Write a paragraph proof.

   Given: X, Y, and Z are noncollinear points.

   Prove: There exists a circle that contains X, Y, and Z.

2. Use a circle to write a paragraph proof for the following theorem, which you learned in Chapter 5:

   The perpendicular bisectors of a triangle intersect at a point that is equidistant from the vertices of a triangle.

3. Write an indirect proof to show that no circle contains three distinct, collinear points.

4. Trace the arc shown at the right. Use a compass and straightedge to construct the center of the circle. Then construct the circle. Explain what you did.

P Q

5. Refer to the diagram. Write a paragraph proof.

   Given: $\overline{IF}$ is a diameter, $\overline{DG} \cong \overline{EH}$.

   Prove: $\angle DJI \cong \angle HJI$

   (*Hint:* Construct perpendicular bisectors.)

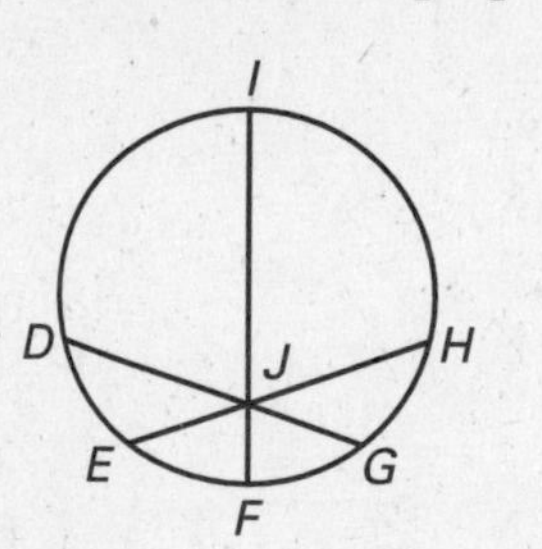

6. Refer to the diagram. If C is the center of the circle, find $m\widehat{DE}$, $m\widehat{EF}$, $m\widehat{FG}$, and $m\widehat{GD}$.

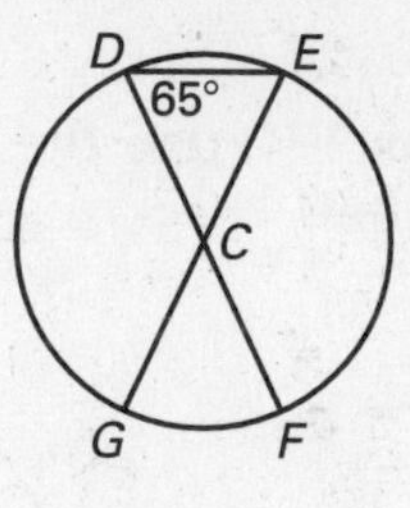

In Exercises 7–9, *C* is the center of the circle. Find the possible values of *x*.

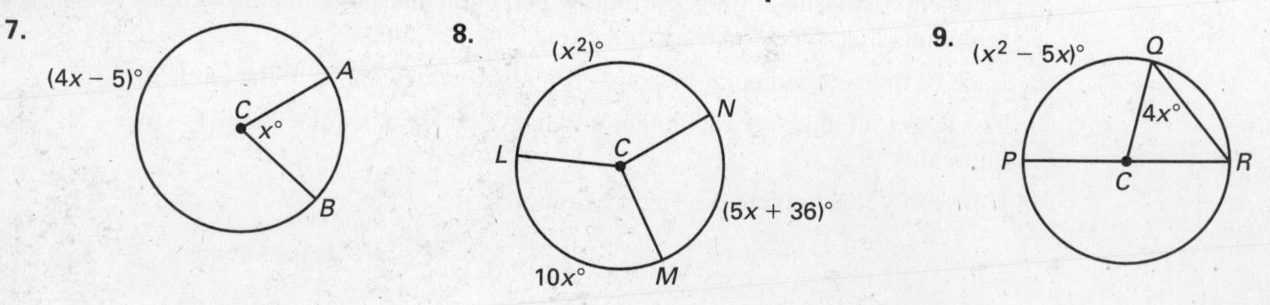

LESSON

10.3

TEACHER'S NAME ____________ CLASS ______ ROOM ______ DATE ______

Lesson Plan

2-day lesson (See *Pacing the Chapter,* TE pages 592C–592D) **For use with pages 612–620**

GOALS

1. Use inscribed angles to solve problems.
2. Use properties of inscribed polygons.

State/Local Objectives ____________

____

✓ Check the items you wish to use for this lesson.

STARTING OPTIONS

____ Homework Check: TE page 607: Answer Transparencies
____ Warm-Up or Daily Homework Quiz: TE pages 613 and 611, CRB page 39, or Transparencies

TEACHING OPTIONS

____ Motivating the Lesson: TE page 614
____ Concept Activity: SE page 612
____ Lesson Opener (Geometry Software): CRB page 40 or Transparencies
____ Technology Activity with Keystrokes: CRB page 41
____ Examples: Day 1: 1–6, SE pages 613–616; Day 2: See the Extra Examples.
____ Extra Examples: Day 1 or Day 2: 1–6, TE pages 614–615 or Transp.
____ Closure Question: TE page 615
____ Guided Practice: SE page 616 Day 1: Exs. 1–8; Day 2: See Checkpoint Exs. TE pages 614–615

APPLY/HOMEWORK

Homework Assignment

____ Basic Day 1: 9–29; Day 2: 30–33, 35–38, 42, 43, 48–58 even, 59–62; Quiz 1: 1–9
____ Average Day 1: 9–29; Day 2: 30–33, 35–38, 41–43, 48–58 even, 59–62; Quiz 1: 1–9
____ Advanced Day 1: 9–29; Day 2: 30–33, 35–47, 48–58 even, 59–62; Quiz 1: 1–9

Reteaching the Lesson

____ Practice Masters: CRB pages 42–44 (Level A, Level B, Level C)
____ Reteaching with Practice: CRB pages 45–46 or Practice Workbook with Examples
____ Personal Student Tutor

Extending the Lesson

____ Applications (Real-Life): CRB page 48
____ Challenge: SE page 619; CRB page 49 or Internet

ASSESSMENT OPTIONS

____ Checkpoint Exercises: Day 1 or Day 2: TE pages 614–615 or Transp.
____ Daily Homework Quiz (10.3): TE page 620, CRB page 53, or Transparencies
____ Standardized Test Practice: SE page 619; TE page 620; STP Workbook; Transparencies
____ Quiz (10.1–10.3): SE page 620; CRB page 50

Notes ____________

____

____

LESSON 10.3

TEACHER'S NAME ______________________ CLASS ________ ROOM ________ DATE ________

Lesson Plan for Block Scheduling

1-day lesson (See *Pacing the Chapter,* TE pages 592C–592D) For use with pages 612–620

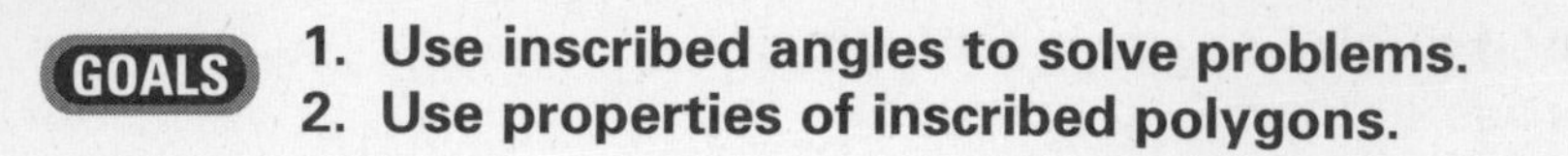

GOALS
1. Use inscribed angles to solve problems.
2. Use properties of inscribed polygons.

State/Local Objectives ____

CHAPTER PACING GUIDE	
Day	**Lesson**
1	10.1 (all)
2	10.2 (all)
3	**10.3 (all)**
4	10.4 (all); 10.5 (begin)
5	10.5 (end); 10.6 (all)
6	10.7 (all)
7	Review Ch. 10; Assess Ch. 10

✓ Check the items you wish to use for this lesson.

STARTING OPTIONS

____ Homework Check: TE page 607: Answer Transparencies
____ Warm-Up or Daily Homework Quiz: TE pages 613 and 611, CRB page 39, or Transparencies

TEACHING OPTIONS

____ Motivating the Lesson: TE page 614
____ Concept Activity: SE page 612
____ Lesson Opener (Geometry Software): CRB page 40 or Transparencies
____ Technology Activity with Keystrokes: CRB page 41
____ Examples 1–6: SE pages 613–616
____ Extra Examples: TE pages 614–615 or Transparencies
____ Closure Question: TE page 615
____ Guided Practice Exercises: SE page 616

APPLY/HOMEWORK

Homework Assignment

____ Block Schedule: 9–33, 35–38, 41–43, 48–58 even, 59–62; Quiz 1: 1–9

Reteaching the Lesson

____ Practice Masters: CRB pages 42–44 (Level A, Level B, Level C)
____ Reteaching with Practice: CRB pages 45–46 or Practice Workbook with Examples
____ Personal Student Tutor

Extending the Lesson

____ Applications (Real-Life): CRB page 48
____ Challenge: SE page 619; CRB page 49 or Internet

ASSESSMENT OPTIONS

____ Checkpoint Exercises: TE pages 614–615 or Transparencies
____ Daily Homework Quiz (10.3): TE page 620, CRB page 53, or Transparencies
____ Standardized Test Practice: SE page 619; TE page 620; STP Workbook; Transparencies
____ Quiz (10.1–10.3): SE page 620; CRB page 50

Notes ____

LESSON

10.3

NAME ______________________ DATE __________

Available as a transparency

WARM-UP EXERCISES

For use before Lesson 10.3, pages 612–620

Solve the equation or system of equations.

1. $x + 3x = 90$
2. $2x + 8x = 180$
3. $2x + 3x + 80 + 120 = 360$
4. $x + y = 60$
   $5x - y = 180$

DAILY HOMEWORK QUIZ

For use after Lesson 10.2, pages 603–611

Determine whether the arc is a *minor arc*, a *major arc*, or a *semicircle* of $\odot K$.

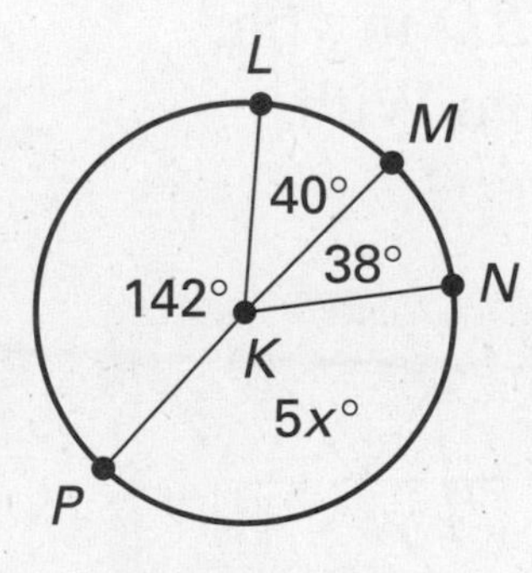

1. $\widehat{PLM}$
2. $\widehat{NL}$

Find the indicated value.

3. $m\widehat{LN}$
4. $m\widehat{PLN}$
5. x

Lesson 10.3

NAME ____________________________ DATE ____________

Available as a transparency

Geometry Software Lesson Opener

For use with pages 613–620

Use geometry software to construct and investigate a "butterfly" inside a circle.

1. Construct a circle and any four points on the circle. Label the points A, B, C, and D, clockwise around the circle. Draw segments AC, BD, AD, and BC. The figure that results is called a *butterfly*.

Drag a point of your butterfly along the circle, without crossing other points, to change its shape. Continue to drag points until you have a butterfly whose shape you like. On paper, sketch and label your butterfly.

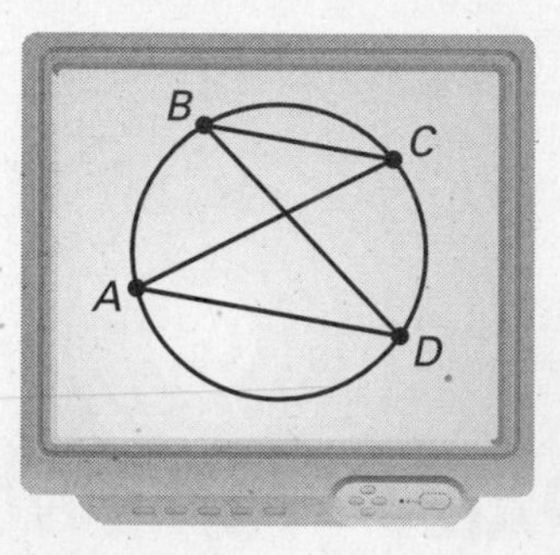

2. Using your butterfly, measure $\angle ADB$ and $\angle BCA$. Drag points and observe how the angle measures change. What do you notice? What relationship does each angle have with arc AB?

3. Predict the relationship between $m\angle DAC$ and $m\angle CBD$ in your butterfly. Measure the angles to check your prediction. With which arc do these angles have a common relationship?

4. Construct a new butterfly with center point M, the midpoint of a chord $\overline{YZ}$ as follows: Start with a large circle and construct any chord $\overline{YZ}$ and its midpoint M. Draw any two chords $\overline{AC}$ and $\overline{BD}$ that contain M. Draw $\overline{AD}$ and $\overline{BC}$, and you have a butterfly. Construct P, the intersection of $\overline{AD}$ and $\overline{YZ}$. Construct Q, the intersection of $\overline{BC}$ and $\overline{YZ}$. Make and test a conjecture about MP and MQ.

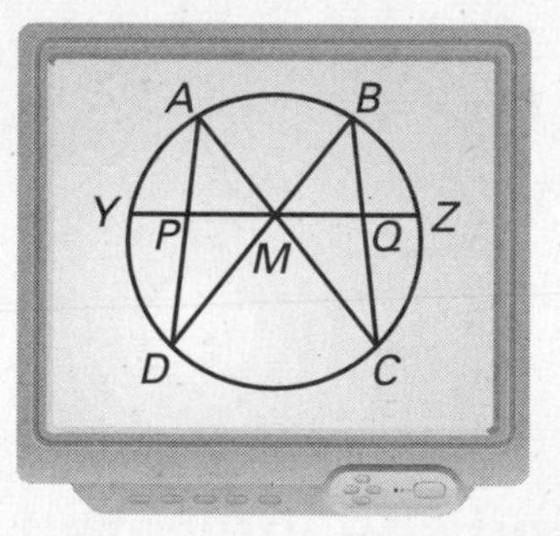

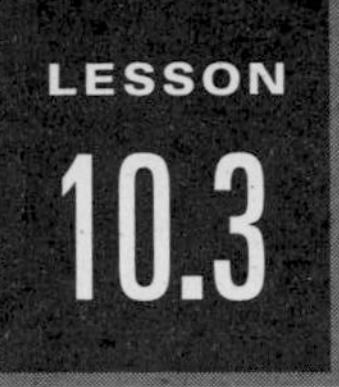

NAME _______________ DATE _______

Technology Activity Keystrokes

For use with page 618

Keystrokes for Exercise 34

TI-92

1. Draw $\odot Q$.

   F3 1 ENTER Q (Move cursor until circle is desired size.) ENTER

2. Draw diameter $\overline{AB}$.

   F2 4 (Move cursor to a point on $\odot Q$.) ENTER (Move cursor center of $\odot Q$.)

   ENTER F2 5 (Move cursor to one point of intersection of this line and $\odot Q$.)

   ENTER A (Move cursor to the other point of intersection of this line and $\odot Q$.)

   ENTER B

3. Plot C on $\odot Q$. F2 2 (Move cursor to a point on $\odot Q$.) ENTER C

4. Draw $\overline{AC}$ and $\overline{CB}$.

   F2 5 (Move cursor to A.) ENTER (Move cursor to C.) ENTER ENTER

   (Move cursor to B.) ENTER

5. Measure the angles of $\triangle ABC$.

   F6 3 (Move cursor to A.) ENTER (Move cursor to C.) ENTER (Move cursor to B.) ENTER F6 3 (Move cursor to B.) ENTER (Move cursor to A.) ENTER (Move cursor to C.) ENTER F6 3 (Move cursor to C.) ENTER (Move cursor to B.) ENTER (Move cursor to A.) ENTER

6. Drag C and notice the results.

   F1 1 (Move cursor to C until prompt says "THIS POINT.") ENTER

   (Use the drag key and the cursor pad to drag the point.)

SKETCHPAD

1. Draw $\odot Q$ using the compass tool.
2. Draw diameter $\overline{AB}$. Choose the line straightedge tool, select a point on $\odot Q$, and draw the line through the center of $\odot Q$. Then choose the segment straightedge tool and draw diameter $\overline{AB}$ by connecting the points of intersection of the line and $\odot Q$.
3. Plot C on $\odot Q$ using the point tool.
4. Draw $\overline{AC}$ and $\overline{CB}$ using the segment straightedge tool.
5. Measure the angles of $\triangle ABC$. For $\angle C$, choose the selection arrow tool, select point A, hold down the shift key and select points C and B (in that order). Choose **Angle** from the **Measure** menu. For $\angle A$, select point B, hold down the shift key and select points A and C (in that order). Choose **Angle** from the **Measure** menu. For $\angle B$, select point A, hold down the shift key and select points B and C (in that order). Choose **Angle** from the **Measure** menu.
6. Drag point C using the translate selection arrow tool and notice the results.

LESSON **10.3**

NAME ______________________ DATE __________

Practice A

For use with pages 613–620

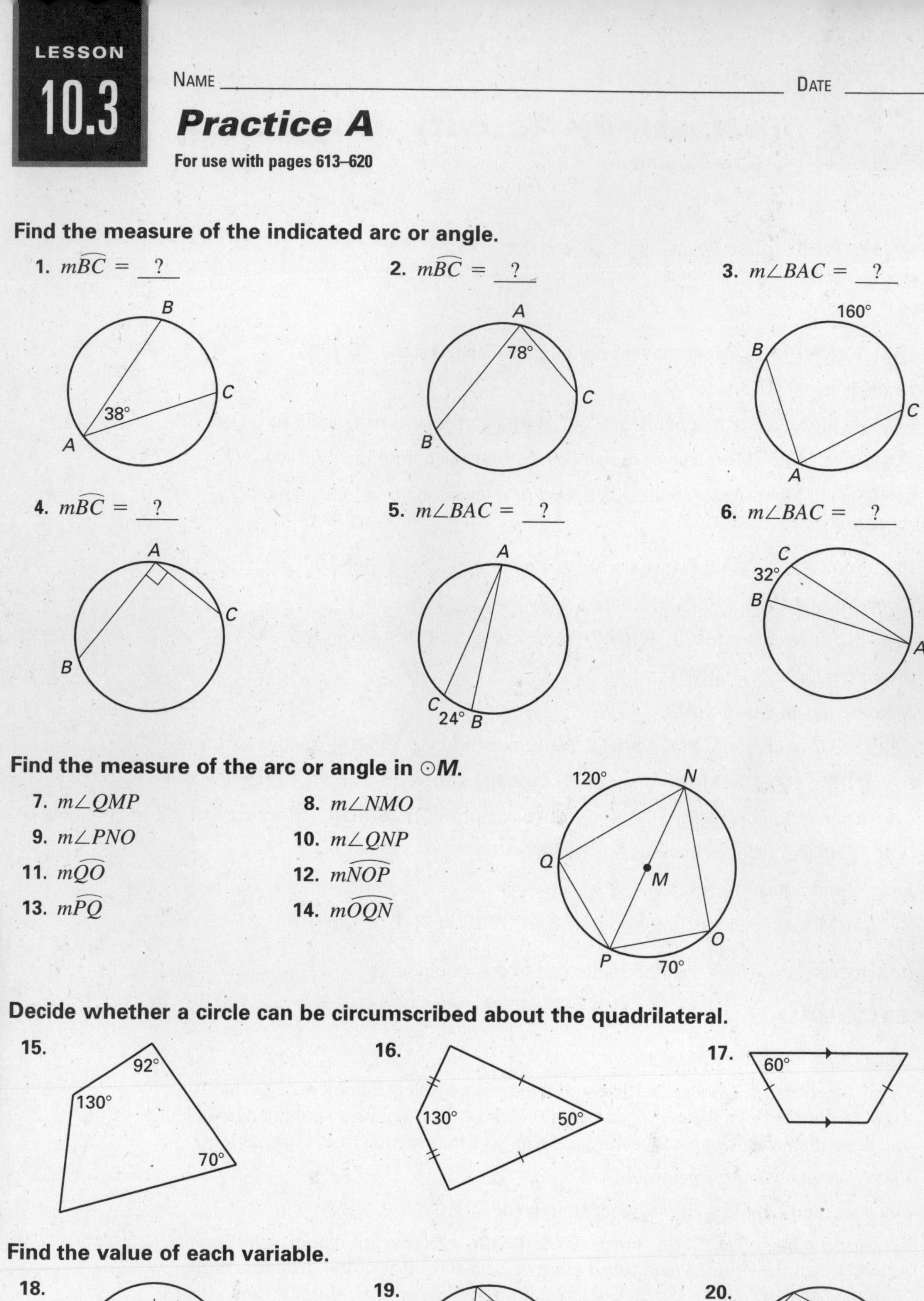

Find the measure of the indicated arc or angle.

1. $m\widehat{BC} = \underline{\ ?\ }$

2. $m\widehat{BC} = \underline{\ ?\ }$

3. $m\angle BAC = \underline{\ ?\ }$

4. $m\widehat{BC} = \underline{\ ?\ }$

5. $m\angle BAC = \underline{\ ?\ }$

6. $m\angle BAC = \underline{\ ?\ }$

Find the measure of the arc or angle in $\odot M$.

7. $m\angle QMP$

8. $m\angle NMO$

9. $m\angle PNO$

10. $m\angle QNP$

11. $m\widehat{QO}$

12. $m\widehat{NOP}$

13. $m\widehat{PQ}$

14. $m\widehat{OQN}$

Decide whether a circle can be circumscribed about the quadrilateral.

15.

16.

17.

Find the value of each variable.

18.

19.

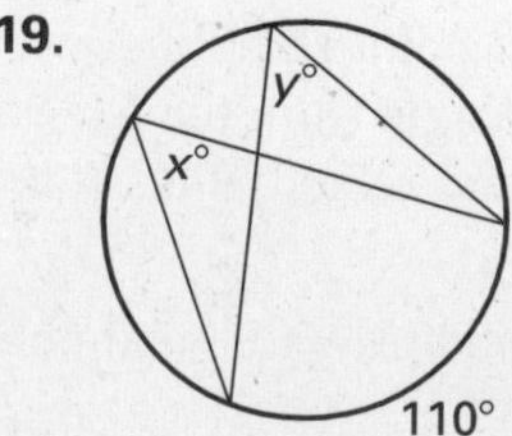

20.

LESSON **10.3**

NAME ______________________________ DATE __________

Practice B

For use with pages 613–620

Find the measure of the indicated arc or angle in $\odot O$.

1. $m\angle BAC = \underline{\ ?\ }$ 2. $m\overarc{BC} = \underline{\ ?\ }$ 3. $m\angle BAC = \underline{\ ?\ }$

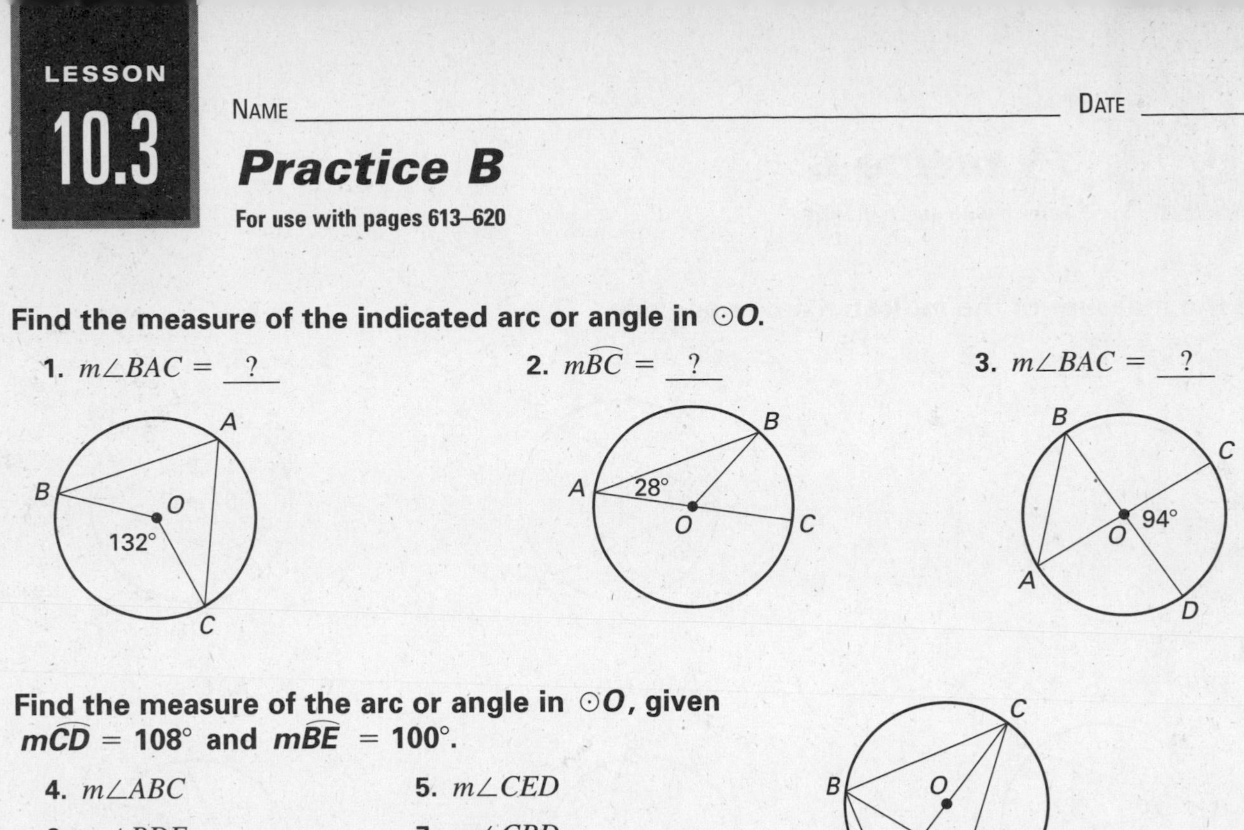

Find the measure of the arc or angle in $\odot O$, given $m\overarc{CD} = 108°$ and $m\overarc{BE} = 100°$.

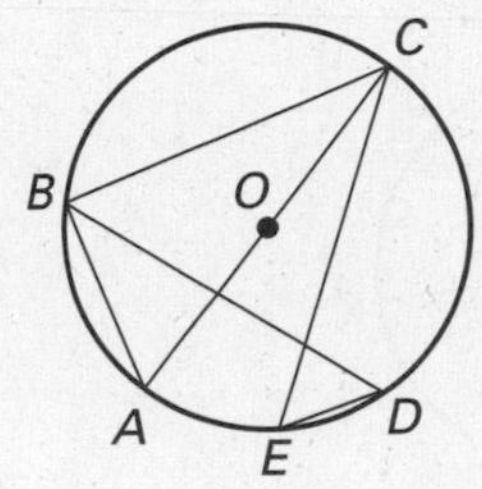

4. $m\angle ABC$
5. $m\angle CED$
6. $m\angle BDE$
7. $m\angle CBD$
8. $m\angle ABD$
9. $m\angle BCE$
10. $m\overarc{AD}$
11. $m\overarc{ABC}$

Find the value of x.

12. 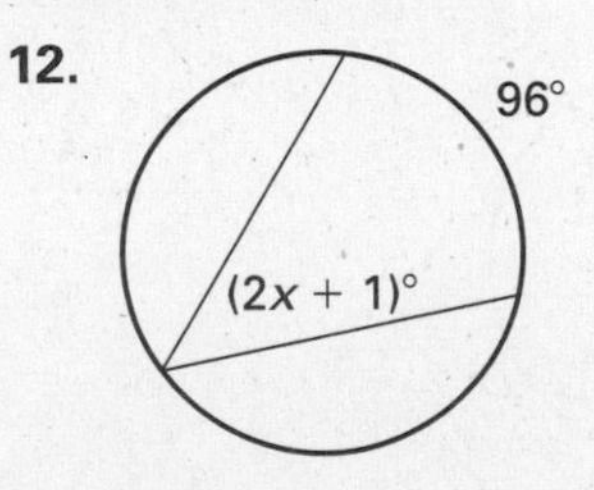

13. 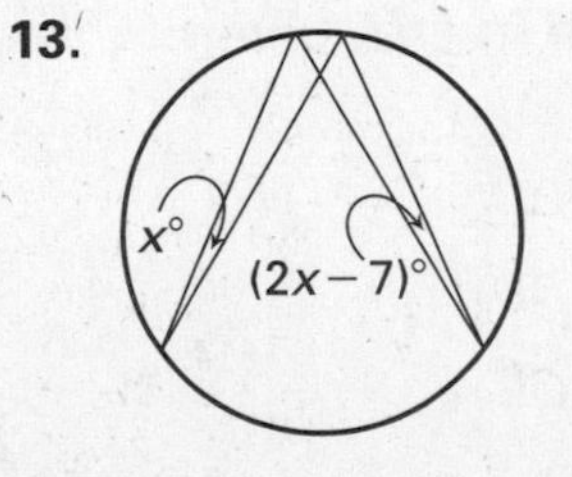

14. diameter $\overline{AB}$

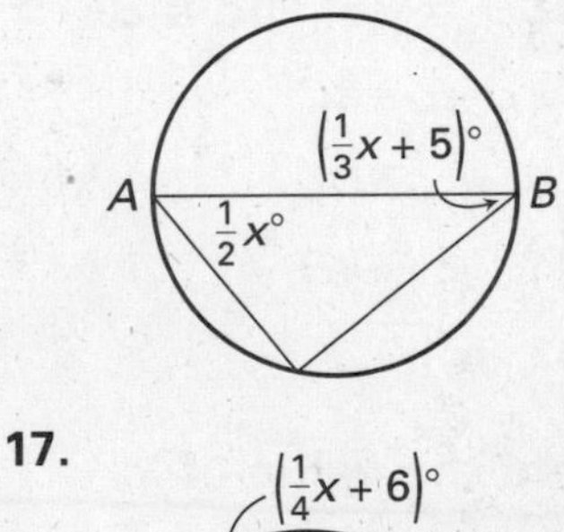

15. 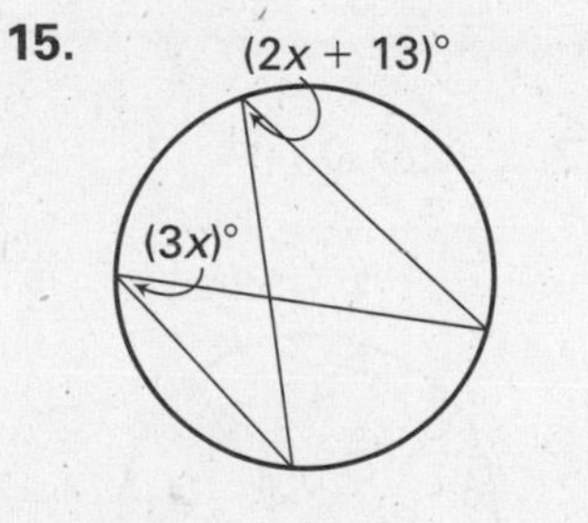

16. 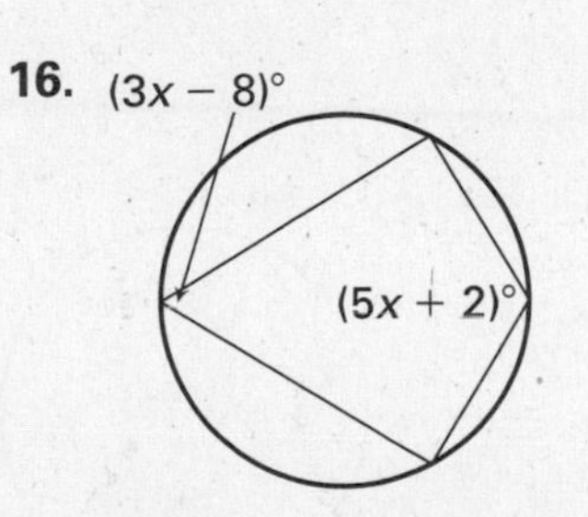

17. 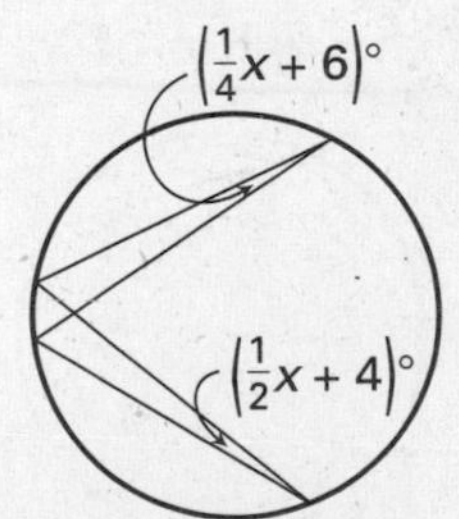

18. ***Archeology*** Archeologists found a portion of a circular dinner plate. Describe a method to determine the diameter of the plate.

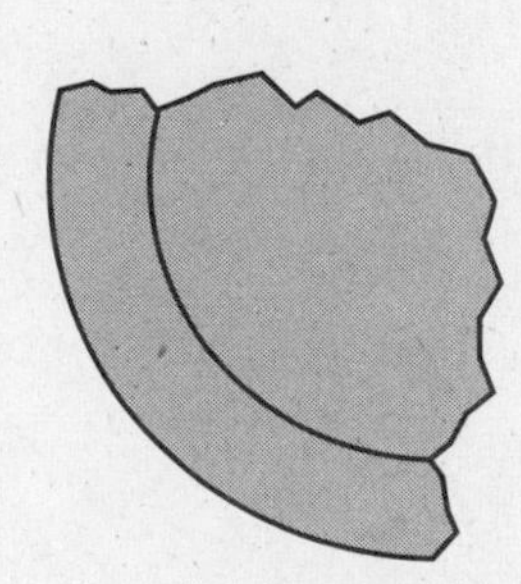

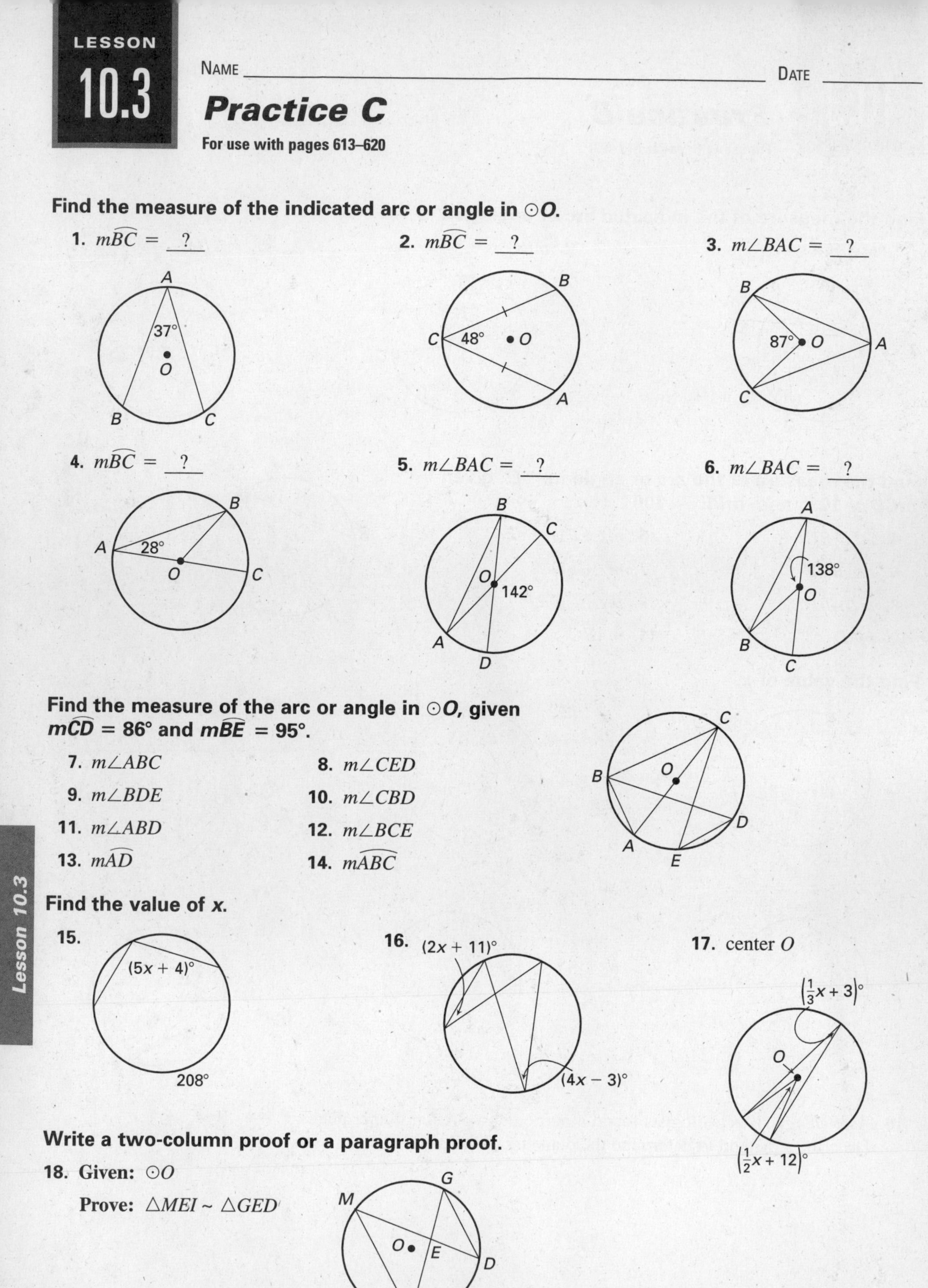

NAME ______________________ DATE __________

Practice C

For use with pages 613–620

Find the measure of the indicated arc or angle in $\odot O$.

1. $m\overset{\frown}{BC} = \underline{\ ?\ }$

2. $m\overset{\frown}{BC} = \underline{\ ?\ }$

3. $m\angle BAC = \underline{\ ?\ }$

4. $m\overset{\frown}{BC} = \underline{\ ?\ }$

5. $m\angle BAC = \underline{\ ?\ }$

6. $m\angle BAC = \underline{\ ?\ }$

Find the measure of the arc or angle in $\odot O$, given $m\overset{\frown}{CD} = 86°$ and $m\overset{\frown}{BE} = 95°$.

7. $m\angle ABC$

8. $m\angle CED$

9. $m\angle BDE$

10. $m\angle CBD$

11. $m\angle ABD$

12. $m\angle BCE$

13. $m\overset{\frown}{AD}$

14. $m\overset{\frown}{ABC}$

Find the value of *x*.

15.

16.

17. center O

Write a two-column proof or a paragraph proof.

18. **Given:** $\odot O$

Prove: $\triangle MEI \sim \triangle GED$

LESSON

NAME ______________________ DATE ____________

Reteaching with Practice

For use with pages 613–620

GOAL **Use inscribed angles to solve problems and use properties of inscribed polygons**

Vocabulary

An **inscribed angle** is an angle whose vertex is on a circle and whose sides contain chords of the circle.

The arc that lies in the interior of an inscribed angle and has endpoints on the angle is called the **intercepted arc** of the angle.

If all of the vertices of a polygon lie on a circle, the polygon is **inscribed** in the circle and the circle is **circumscribed** about the polygon.

Theorem 10.8 Measure of an Inscribed Angle
If an angle is inscribed in a circle, then its measure is half the measure of its intercepted arc.

Theorem 10.9
If two inscribed angles of a circle intercept the same arc, then the angles are congruent.

Theorem 10.11
A quadrilateral can be inscribed in a circle if and only if its opposite angles are supplementary.

EXAMPLE 1 *Finding Measures of Arcs and Inscribed Angles*

Find the value of x.

a.

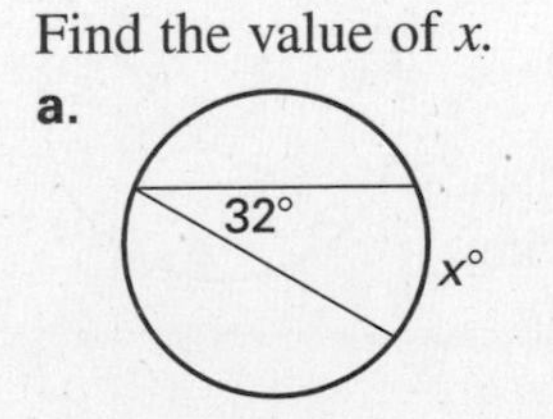

b.

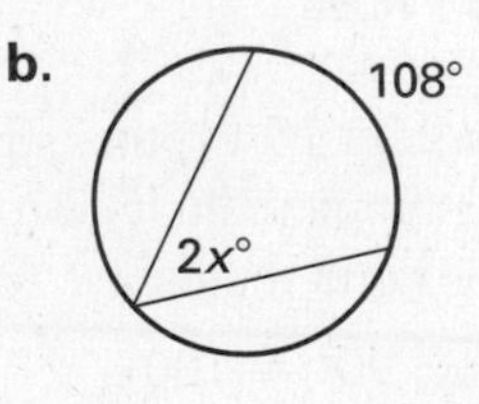

Solution

a. By Theorem 10.8,

$$32° = \frac{1}{2}x°$$

$$64 = x$$

b. $2x° = \frac{1}{2}(108°)$

$$2x = 54$$

$$x = 27$$

Exercises for Example 1

Find the value of *x*.

1.

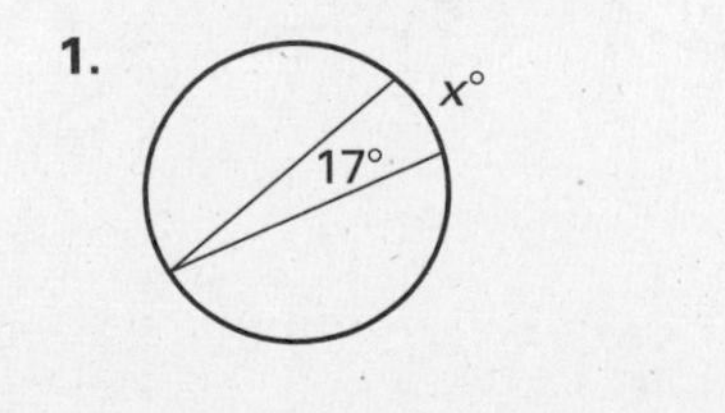

2.

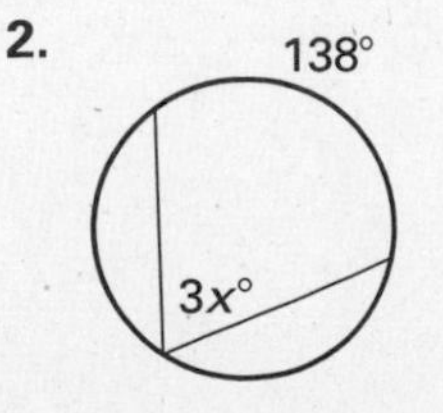

3.

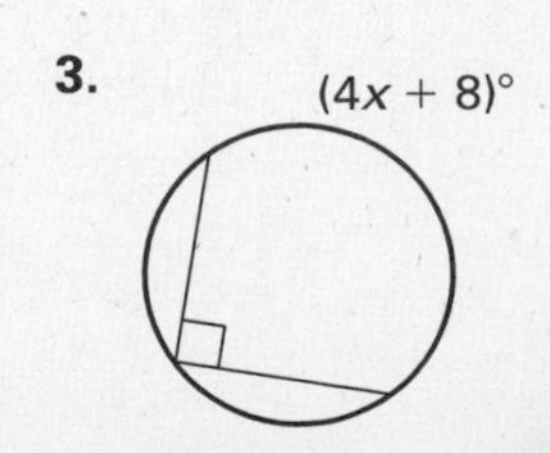

NAME ______________________________ DATE ____________

Reteaching with Practice

For use with pages 613–620

EXAMPLE 2 Finding the Measure of an Angle

If $\angle CAD$ is a right angle, what is the measure of $\angle CBD$?

SOLUTION

By Theorem 10.9, $\angle CAD \cong \angle CBD$ because the two angles intercept the same arc. So, $m\angle CBD = 90°$.

Exercises for Example 2

Find the value of *x*.

4.

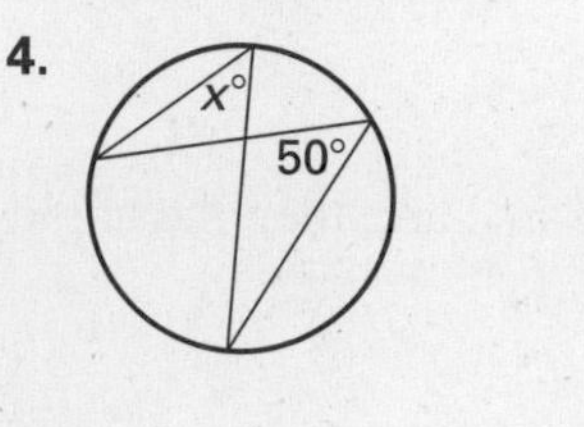

5.

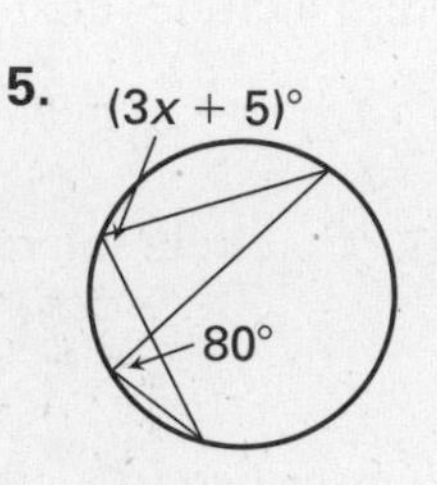

6.

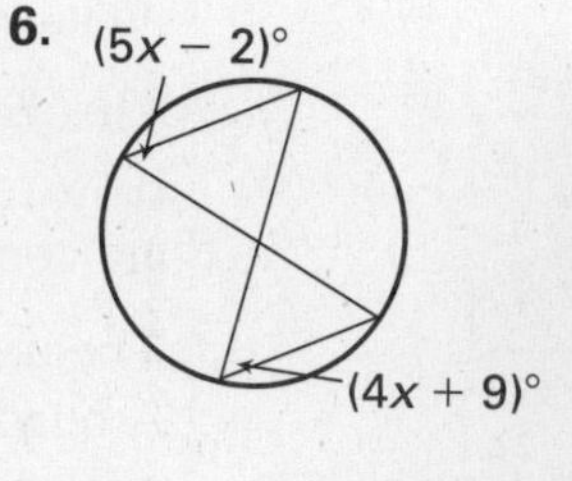

EXAMPLE 3 Using an Inscribed Quadrilateral

Find the value of each variable.

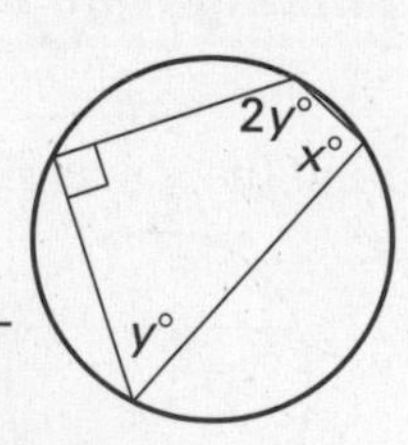

SOLUTION

By Theorem 10.11, the opposite angles of this quadrilateral are supplementary. So you can write the following equations and then solve for the variable in each.

$x° + 90° = 180°$ $\qquad$ $2y° + y° = 180°$

$x = 90$ $\qquad$ $3y = 180$

$y = 60$

Exercises for Example 3

Find the value of each variable.

7. **8.**

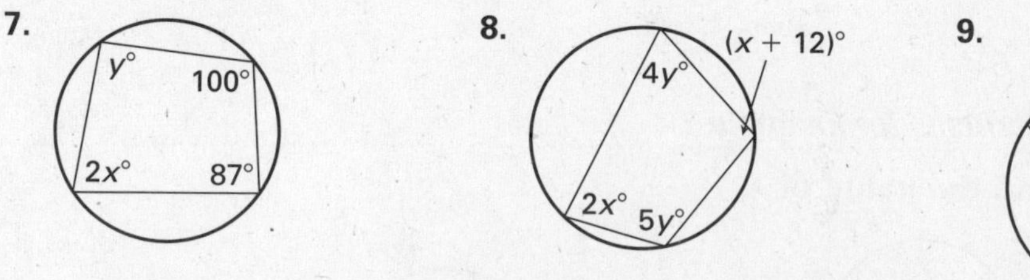

9.

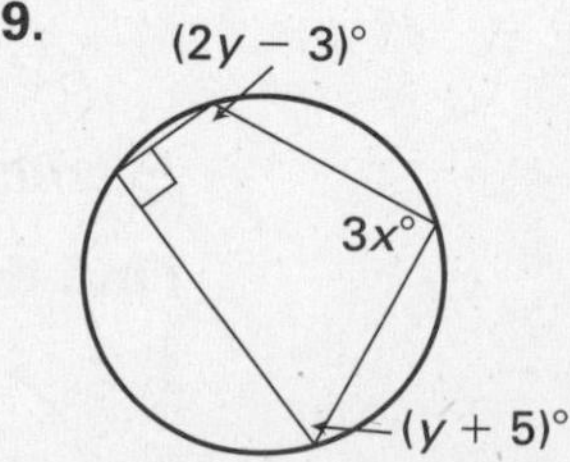

NAME ______________________________ DATE __________

Quick Catch-Up for Absent Students

For use with pages 612–620

The items checked below were covered in class on (date missed) ___________

Activity 10.3: Investigating Inscribed Angles (p. 612)

____ **Goal:** Determine how the measure of an inscribed angle relates to the measure of its corresponding central angle.

Lesson 10.3: Inscribed Angles

____ **Goal 1:** Use inscribed angles to solve problems. (pp. 613–614)

Material Covered:

____ Example 1: Finding Measures of Arcs and Inscribed Angles

____ Example 2: Comparing Measures of Inscribed Angles

____ Example 3: Finding the Measure of an Angle

____ Example 4: Using the Measure of an Inscribed Angle

Vocabulary:

inscribed angle, p. 613 intercepted arc, p. 613

____ **Goal 2:** Use properties of inscribed polygons. (pp. 615–616)

Material Covered:

____ Example 5: Using Theorems 10.10 and 10.11

____ Student Help: Look Back

____ Example 6: Using an Inscribed Quadrilateral

Vocabulary:

inscribed, p. 615 circumscribed, p. 615

____ Other (specify) ______________________________

____

Homework and Additional Learning Support

____ Textbook (specify) pp. 617–620

____

____ *Reteaching with Practice* worksheet (specify exercises) ______________

____ *Personal Student Tutor* for Lesson 10.3

NAME ______________________________ DATE __________

Real-Life Application: When Will I Ever Use This?

For use with pages 613–620

Satellite Television

Satellite television companies use stationary satellites to transmit programming to their customers. They need the stable coverage to ensure customer satisfaction. The companies must maximize the area covered by each satellite in their network. They want to launch as few satellites as possible to completely cover an area. To do this they must find out how much of Earth is covered by one satellite.

In Exercises 1–6, use the following information.

The satellite at point A is a stationary satellite in the network of a satellite television company. It is hovering 400 miles above Earth. The radius of Earth is approximately 4000 miles.

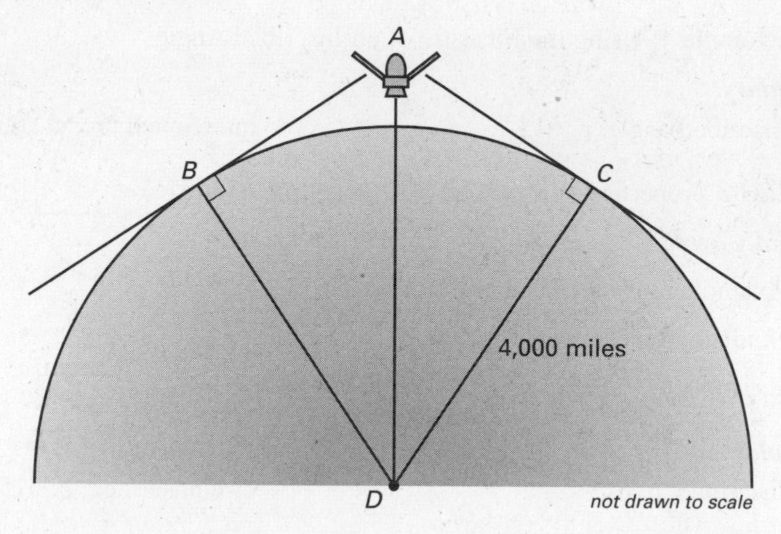

1. Name the central angle that intercepts the arc covered by the satellite.
2. What is the length of $\overline{AD}$?
3. Since $\triangle ADB \cong \triangle ADC$ (proved in the Real-Life Application in 10.1), the central angle is two times $\angle ADC$. Using an inverse trig function, find $\angle ADC$ to the nearest degree. (Hint: You should know AD and CD.)
4. What is the measure of the central angle?
5. How many of these satellites would it take to cover the whole circumference of Earth? (Hint: How many degrees do you have to cover?)
6. How far apart would the satellites be?

LESSON 10.3

NAME ______________________________ DATE __________

Challenge: Skills and Applications

For use with pages 613–620

1. Refer to the diagram. Write a paragraph proof.

   Given: $\odot C$, $\overline{FG} \cong \overline{GE}$

   Prove: $\triangle DEF$ is isosceles.

   (*Hint:* Draw an additional segment.)

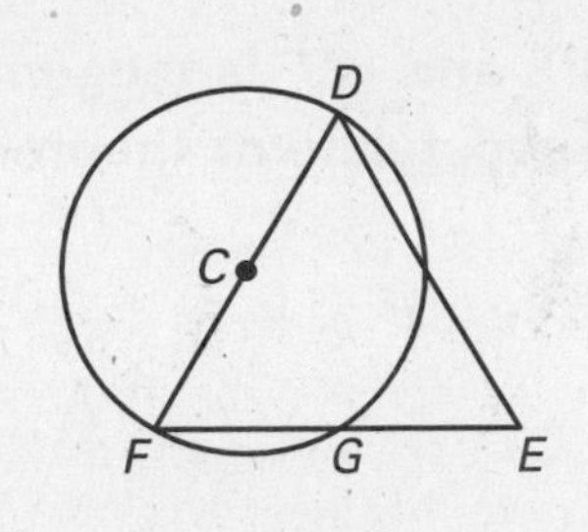

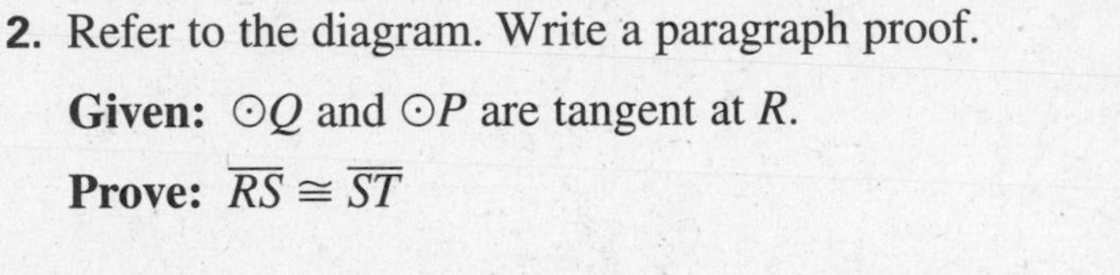

2. Refer to the diagram. Write a paragraph proof.

   Given: $\odot Q$ and $\odot P$ are tangent at R.

   Prove: $\overline{RS} \cong \overline{ST}$

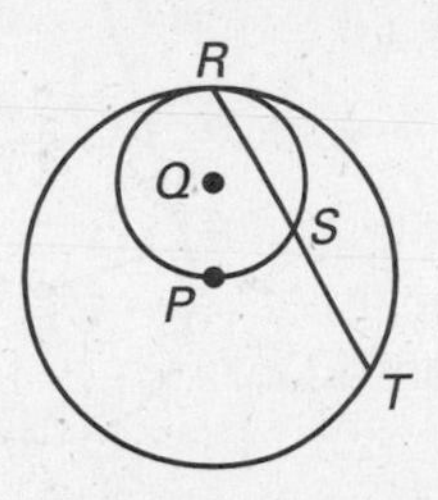

3. Refer to the diagram. Write a paragraph proof.

   Given: $\overline{WZ}$ is a base of trapezoid $WXYZ$ inscribed in $\odot V$.

   Prove: $WXYZ$ is an isosceles trapezoid.

4. In this exercise, you will use a circle to prove the Pythagorean Theorem. Given right triangle $\triangle OPQ$ with side lengths a, b, and c, as shown, draw a circle centered at O with radius c. Let R and S be the points where $\overleftrightarrow{OP}$ intersects the circle, as shown.

   a. Find lengths OR and PS in terms of a, b, and c.

   b. What kind of triangle is $\triangle QRS$?

   c. Use a Geometric Mean Theorem to write an equation involving PQ.

   d. Now show that $a^2 + b^2 = c^2$.

Lesson 10.3

LESSON

10.3

NAME ____________________ DATE __________

Quiz 1

For use after Lessons 10.1–10.3

$\overleftrightarrow{WR}$ is tangent to $\odot S$ at R and $\overleftrightarrow{WT}$ is tangent to $\odot S$ at T. Find the value of x. Write the postulate or theorem that justifies your answer. *(Lesson 10.1)*

1. **2.**

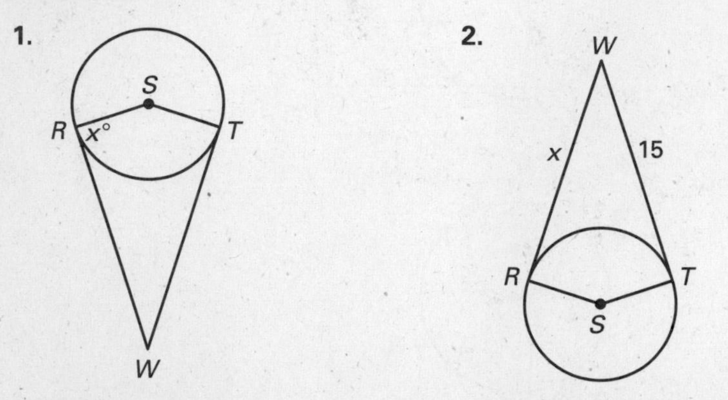

Answers

1. ____________
2. ____________
3. ____________
4. ____________
5. ____________
6. ____________
7. ____________
8. ____________
9. ____________

Find the measure of the arc of $\odot P$. *(Lesson 10.2)*

3. $\overset{\frown}{XY}$

4. $\overset{\frown}{ZW}$

5. $\overset{\frown}{XYW}$

6. $\overset{\frown}{XYZ}$

7. $\overset{\frown}{YW}$

8. $\overset{\frown}{ZYW}$

X Y 120° P Z W

9. If an angle that has a measure of 51.4° is inscribed in a circle, what is the measure of its intercepted arc? *(Lesson 10.3)*

Lesson 10.3

LESSON 10.4

TEACHER'S NAME ______________________ CLASS ______ ROOM ______ DATE ______

Lesson Plan

1-day lesson (See *Pacing the Chapter,* TE pages 592C–592D) For use with pages 621–627

GOALS

1. **Use angles formed by tangents and chords to solve problems in geometry.**
2. **Use angles formed by lines that intersect a circle to solve problems.**

State/Local Objectives ____

____

✓ Check the items you wish to use for this lesson.

STARTING OPTIONS

____ Homework Check: TE page 616: Answer Transparencies
____ Warm-Up or Daily Homework Quiz: TE pages 621 and 620, CRB page 53, or Transparencies

TEACHING OPTIONS

____ Lesson Opener (Visual Approach): CRB page 54 or Transparencies
____ Technology Activity with Keystrokes: CRB page 55
____ Examples 1–5: SE pages 621–623
____ Extra Examples: TE pages 622–623 or Transparencies; Internet
____ Closure Question: TE page 623
____ Guided Practice Exercises: SE page 624

APPLY/HOMEWORK

Homework Assignment

____ Basic 8–34, 40, 42, 43, 46–51
____ Average 8–35, 40, 42, 43, 46–51
____ Advanced 8–35, 37–40, 42–51

Reteaching the Lesson

____ Practice Masters: CRB pages 56–58 (Level A, Level B, Level C)
____ Reteaching with Practice: CRB pages 59–60 or Practice Workbook with Examples
____ Personal Student Tutor

Extending the Lesson

____ Applications (Interdisciplinary): CRB page 62
____ Challenge: SE page 627; CRB page 63 or Internet

ASSESSMENT OPTIONS

____ Checkpoint Exercises: TE pages 622–623 or Transparencies
____ Daily Homework Quiz (10.4): TE page 627, CRB page 66, or Transparencies
____ Standardized Test Practice: SE page 627; TE page 627; STP Workbook; Transparencies

Notes ____

____

____

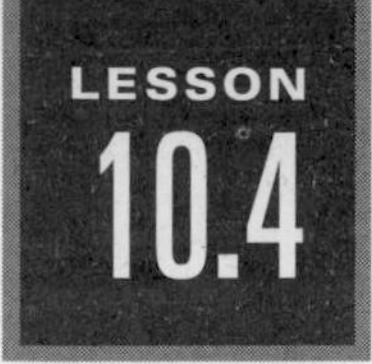

LESSON 10.4

TEACHER'S NAME ______________________ CLASS _______ ROOM _______ DATE _______

Lesson Plan for Block Scheduling

Half-day lesson (See *Pacing the Chapter,* TE pages 592C–592D) **For use with pages 621–627**

GOALS

1. **Use angles formed by tangents and chords to solve problems in geometry.**
2. **Use angles formed by lines that intersect a circle to solve problems.**

State/Local Objectives ____

CHAPTER PACING GUIDE	
Day	**Lesson**
1	10.1 (all)
2	10.2 (all)
3	10.3 (all)
4	**10.4 (all)**; 10.5 (begin)
5	10.5 (end); 10.6 (all)
6	10.7 (all)
7	Review Ch. 10; Assess Ch. 10

✓ Check the items you wish to use for this lesson.

STARTING OPTIONS

____ Homework Check: TE page 616: Answer Transparencies
____ Warm-Up or Daily Homework Quiz: TE pages 621 and 620, CRB page 53, or Transparencies

TEACHING OPTIONS

____ Lesson Opener (Visual Approach): CRB page 54 or Transparencies
____ Technology Activity with Keystrokes: CRB page 55
____ Examples 1–5: SE pages 621–623
____ Extra Examples: TE pages 622–623 or Transparencies; Internet
____ Closure Question: TE page 623
____ Guided Practice Exercises: SE page 624

APPLY/HOMEWORK

Homework Assignment (See also the assignment for Lesson 10.5.)

____ Block Schedule: 8–35, 40, 42, 43, 46–51

Reteaching the Lesson

____ Practice Masters: CRB pages 56–58 (Level A, Level B, Level C)
____ Reteaching with Practice: CRB pages 59–60 or Practice Workbook with Examples
____ Personal Student Tutor

Extending the Lesson

____ Applications (Interdisciplinary): CRB page 62
____ Challenge: SE page 627; CRB page 63 or Internet

ASSESSMENT OPTIONS

____ Checkpoint Exercises: TE pages 622–623 or Transparencies
____ Daily Homework Quiz (10.4): TE page 627, CRB page 66, or Transparencies
____ Standardized Test Practice: SE page 627; TE page 627; STP Workbook; Transparencies

Notes ____

LESSON

10.4

NAME ______________________ DATE __________

Available as a transparency

WARM-UP EXERCISES

For use before Lesson 10.4, pages 621–627

Solve the equation.

1. $4c = 180$

2. $\frac{1}{2}(3x + 42) = 27$

3. $8y = \frac{1}{2}(5y + 55)$

4. $120 = \frac{1}{2}[(360 - x) - x]$

DAILY HOMEWORK QUIZ

For use after Lesson 10.3, pages 612–620

Find the value of each variable.

1.

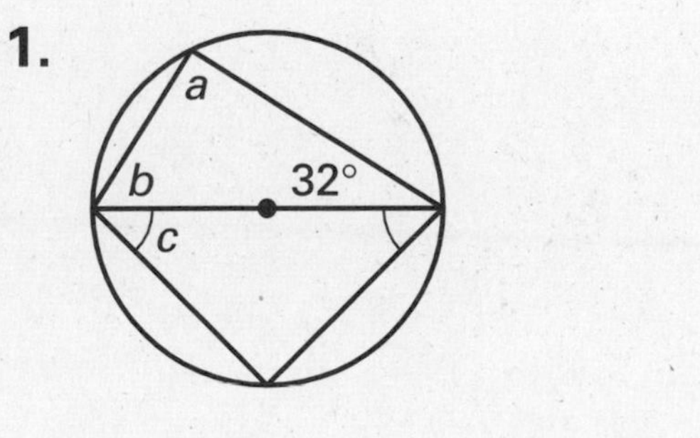

2.

(15y + 3)°
10(y + 4)°
7x°
(8x + 2)°

Lesson 10.4

NAME ________________________________ DATE __________

Available as a transparency

Visual Approach Lesson Opener

For use with pages 621–627

Each diagram shows two intersecting lines that intersect a circle.

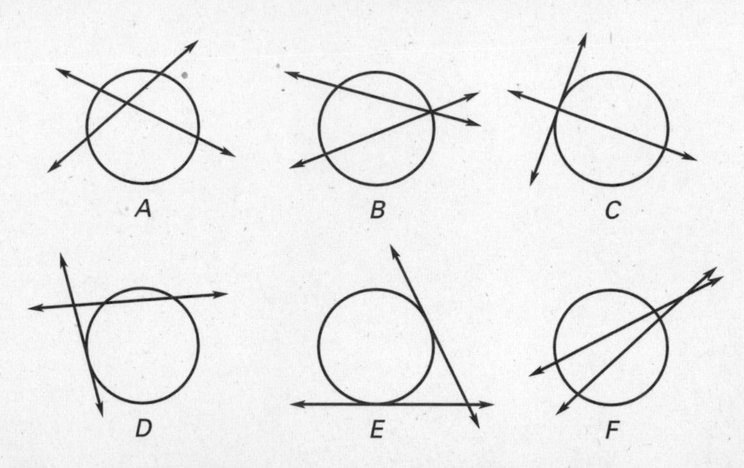

Put the diagrams into categories by writing the letter of each diagram that answers the question in the given way.

1. Where is the point of intersection of the two lines?

a. inside the circle

b. outside the circle

c. on the circle

2. What types of lines intersect the circle?

a. two secants

b. two tangents

c. one secant and one tangent

3. How many points of intersection are there between the circle and the lines?

a. two points

b. three points

c. four points

LESSON 10.4

NAME ______________________________ DATE __________

Reteaching with Practice

For use with pages 621–627

GOAL **Use angles formed by tangents and chords to solve problems in geometry and use angles formed by lines that intersect a circle to solve problems**

VOCABULARY

Theorem 10.12
If a tangent and a chord intersect at a point on a circle, then the measure of each angle formed is one half the measure of its intercepted arc.

Theorem 10.13
If two chords intersect in the *interior* of a circle, then the measure of each angle is one half the *sum* of the measures of the arcs intercepted by the angle and its vertical angle.

Theorem 10.14
If a tangent and a secant, two tangents, or two secants intersect in the *exterior* of a circle, then the measure of the angle formed is one half the *difference* of the measures of the intercepted arcs.

EXAMPLE 1 Finding Angle and Arc Measures

Line m is tangent to the circle.

a. Find $m\angle 1$

b. $m\overset{\frown}{ACB}$

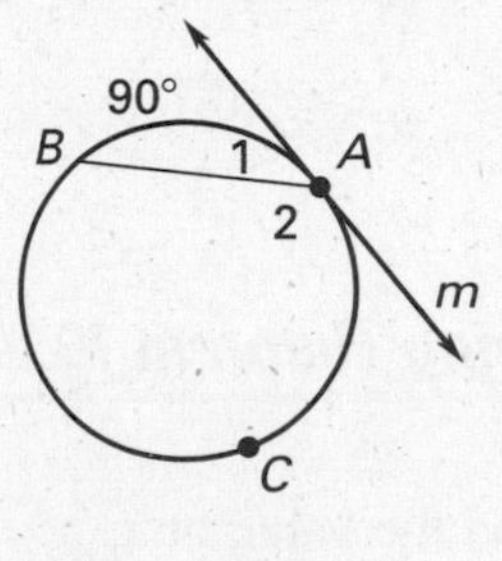

SOLUTION

a. $m\angle 1 = \frac{1}{2}(90°) = 45°$

b. Because $\angle 1$ and $\angle 2$ are a linear pair,
$m\angle 2 = 180° - m\angle 1 = 180° - 45° = 135°$. So,
$m\overset{\frown}{ACB} = 2(135°) = 270°$.

Exercises for Example 1

Find the value of each variable.

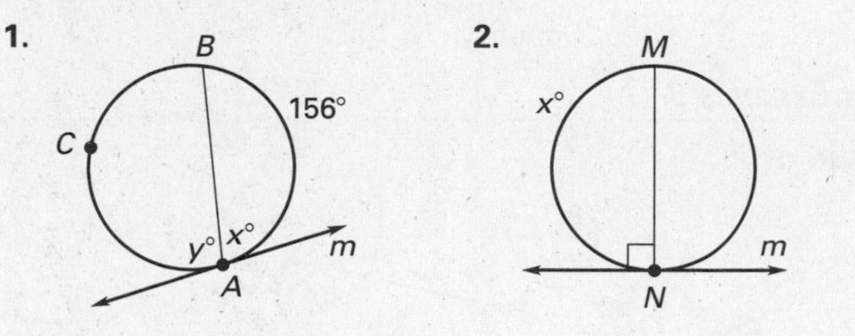

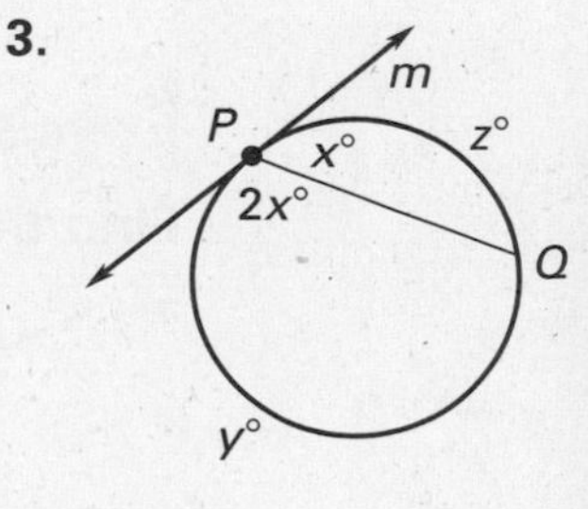

Lesson 10.4

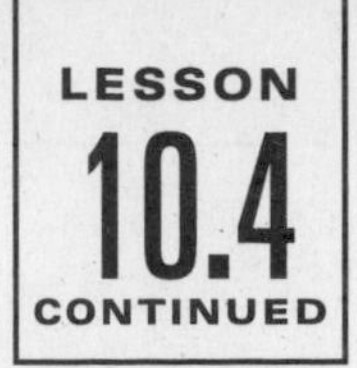

NAME ______________________________ DATE __________

Reteaching with Practice

For use with pages 621–627

EXAMPLE 2 Using Theorem 10.13

Find the value of x.

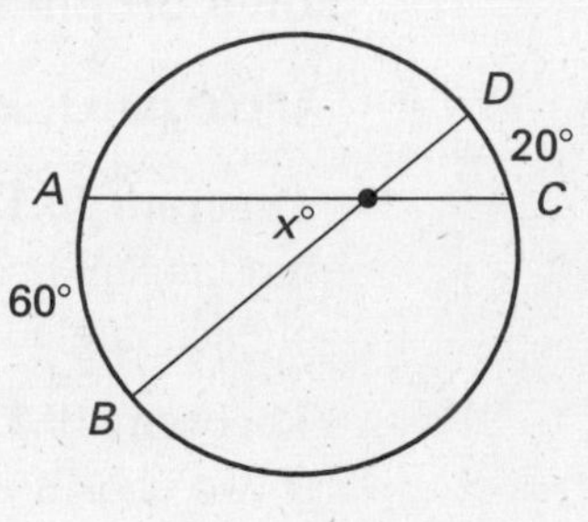

SOLUTION

$x° = \frac{1}{2}(m\widehat{AB} + m\widehat{CD})$ Apply Theorem 10.13.

$x° = \frac{1}{2}(60° + 20°)$ Substitute.

$x = 40$ Simplify.

Exercises for Example 2

Find the value of x.

4.

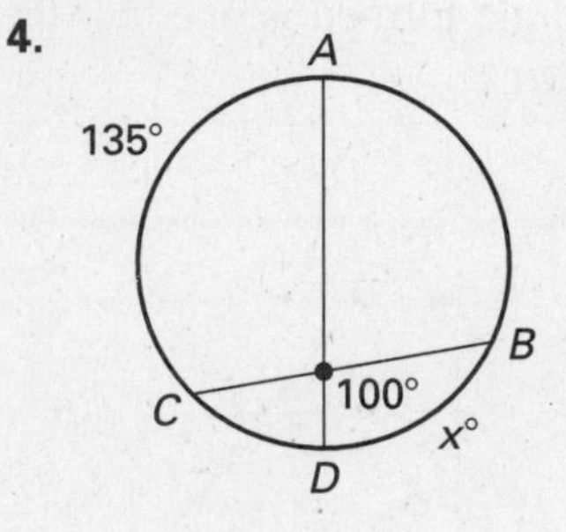

5.

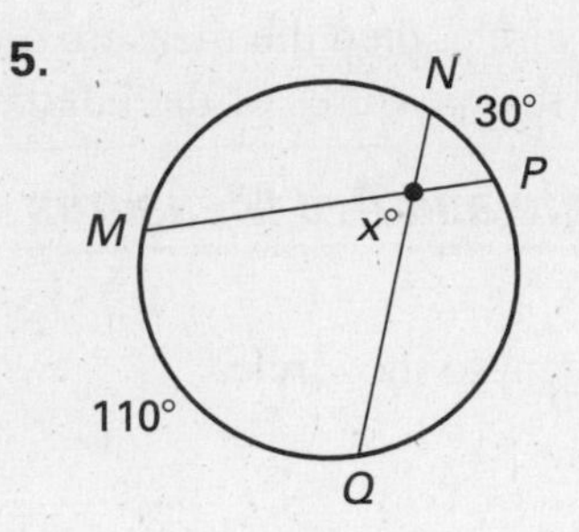

6.

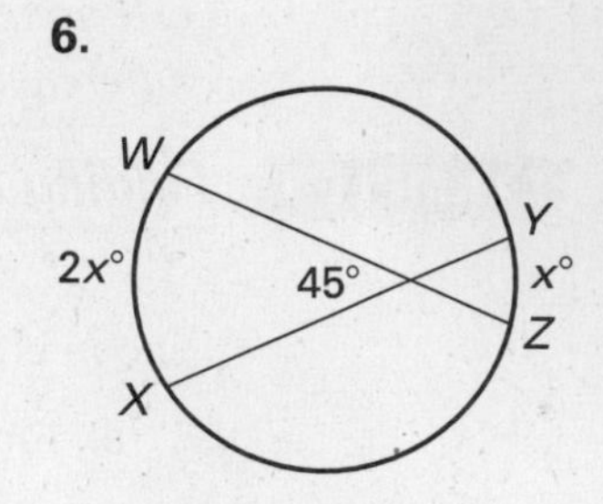

EXAMPLE 3 Using Theorem 10.14

Find the value of x.

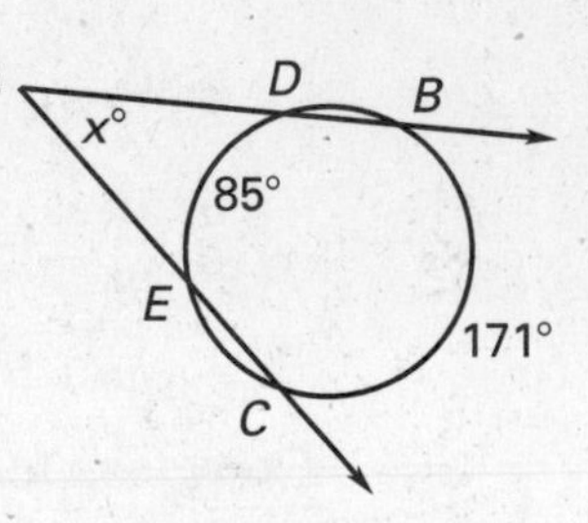

SOLUTION

$x° = \frac{1}{2}(m\widehat{BC} - m\widehat{DE})$ Apply Theorem 10.14.

$x° = \frac{1}{2}(171° - 85°)$ Substitute.

$x = 43$ Simplify.

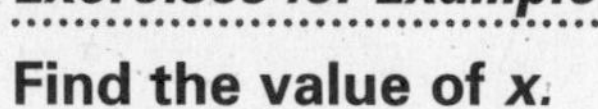

Exercises for Example 3

Find the value of x.

7.

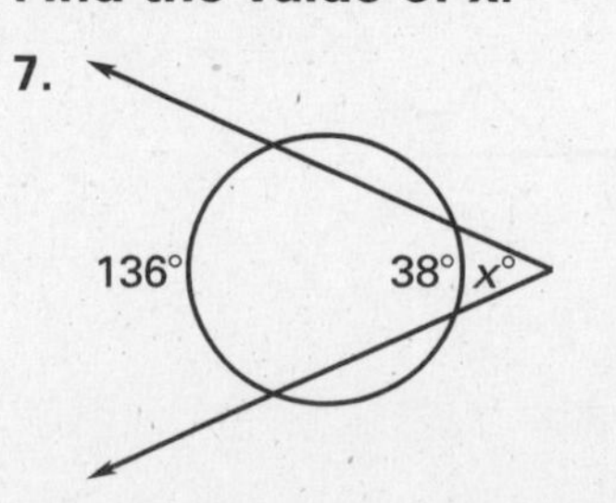

8.

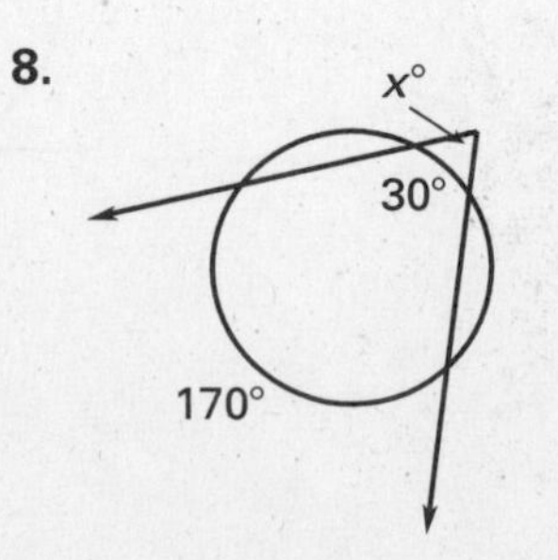

9.

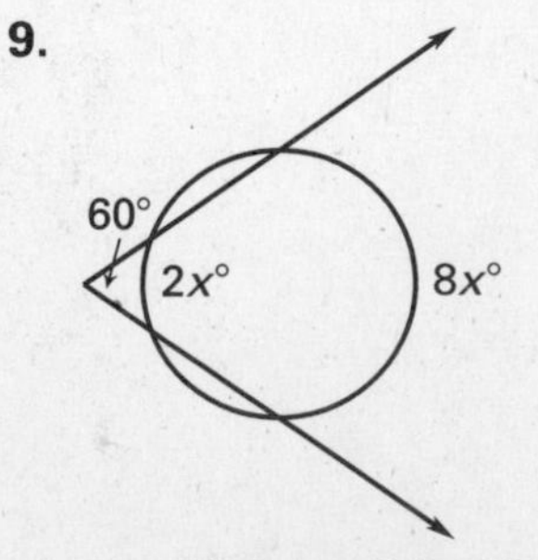

Lesson 10.4

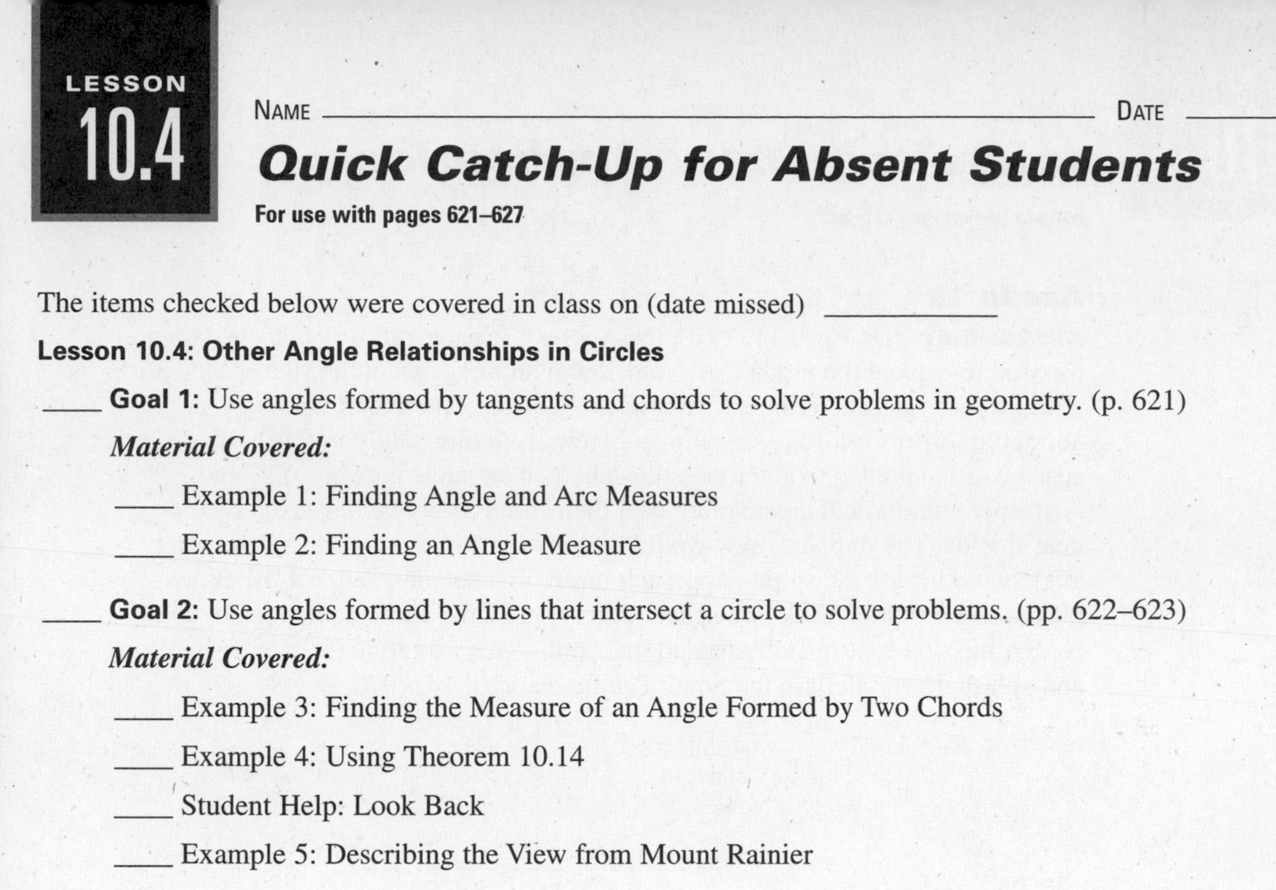

LESSON 10.4

NAME ______________________ DATE __________

Quick Catch-Up for Absent Students

For use with pages 621–627

The items checked below were covered in class on (date missed) __________

Lesson 10.4: Other Angle Relationships in Circles

____ **Goal 1:** Use angles formed by tangents and chords to solve problems in geometry. (p. 621)

Material Covered:

____ Example 1: Finding Angle and Arc Measures

____ Example 2: Finding an Angle Measure

____ **Goal 2:** Use angles formed by lines that intersect a circle to solve problems. (pp. 622–623)

Material Covered:

____ Example 3: Finding the Measure of an Angle Formed by Two Chords

____ Example 4: Using Theorem 10.14

____ Student Help: Look Back

____ Example 5: Describing the View from Mount Rainier

____ Other (specify) ______________________

____

Homework and Additional Learning Support

____ Textbook (specify) pp. 624–627

____

____ Internet: Extra Examples at www.mcdougallittell.com

____ *Reteaching with Practice* worksheet (specify exercises) __________

____ *Personal Student Tutor* for Lesson 10.4

Lesson 10.4

NAME ____________________ DATE ____________

Interdisciplinary Application

For use with pages 621–627

Apollo 13

ASTRONOMY On April 11, 1970, the Apollo 13 spacecraft was launched on a mission to explore the moon. An explosion in an oxygen tank left the spacecraft with little oxygen and crippled navigational devices. The moon landing was aborted in order to salvage the ship and crew. To return safely to Earth, the space capsule needed to enter the atmosphere at an angle between 5.5° and 7.5°. A steeper angle would create more heat than could safely be dispersed by the heat shields. The ship and crew would be consumed in a fiery crash. A smaller angle would cause the ship to approach Earth on a tangent, skip off the atmosphere, and be hurled back into space. Careful planning and calculations by the NASA mission control staff enabled the Apollo crew to refine the reentry angle and splash down safely in the South Pacific on April 17, 1970.

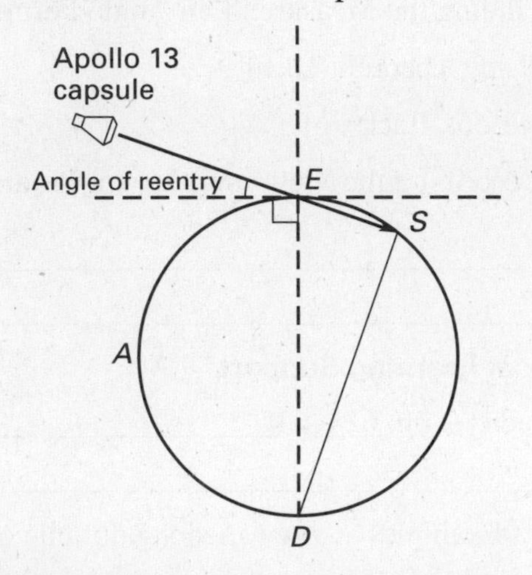

In Exercises 1–5, use the diagram above where $\overline{ED}$ is the diameter of Earth.

1. Find the measure of $\angle ESD$.
2. Find the measure of $\overset{\frown}{DAE}$.
3. Find the measure of $\overset{\frown}{ES}$ if the measure of $\overset{\frown}{SD}$ is 167°.
4. Find the measure of $\angle EDS$ if the measure of $\overset{\frown}{ES}$ is 13°.
5. Find the angle of reentry for the Apollo 13 Space Capsule.

NAME ______________________________ DATE ____________

Challenge: Skills and Applications

For use with pages 621–627

1. Refer to the diagram. Write a paragraph proof.

Given: The two circles are tangent at T, line k is tangent to the smaller circle at U, line j is tangent to the larger circle at V.

Prove: $j \parallel k$

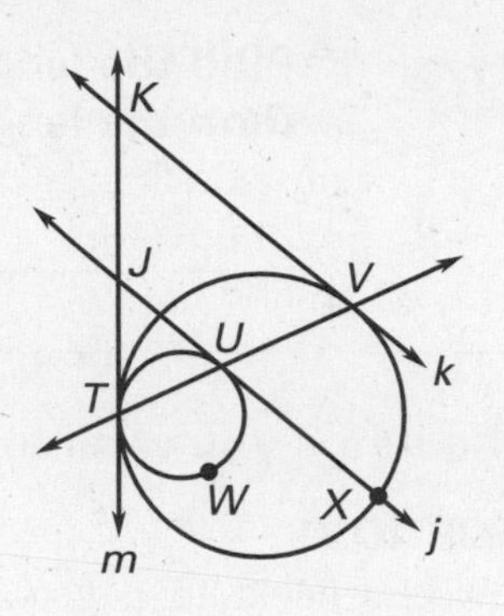

2. Refer to the diagram. Write a paragraph proof.

Given: $m\widehat{AG} = 2m\widehat{CE}$

Prove: $m\widehat{CE} + 2m\widehat{BF} = 360°$

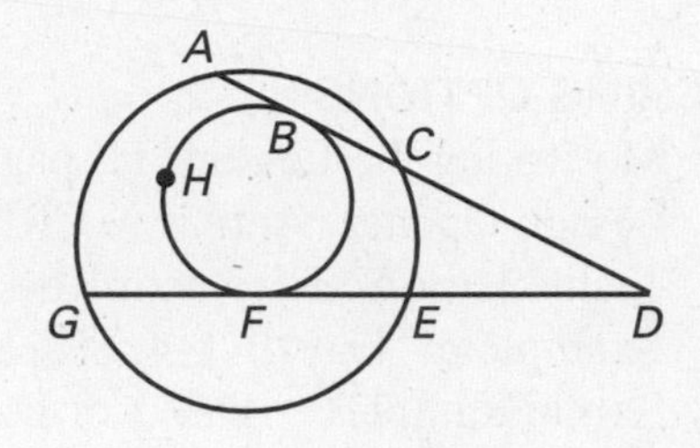

3. Refer to the diagram. Write an expression for $m\angle P$ in terms of x and y.

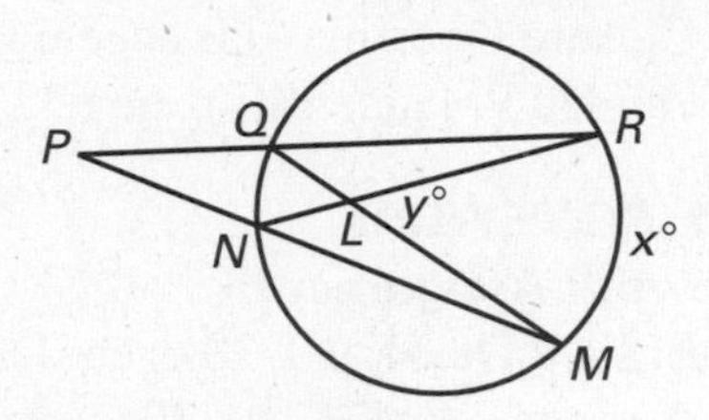

In Exercises 4–9, find the possible values of *x*.

4.

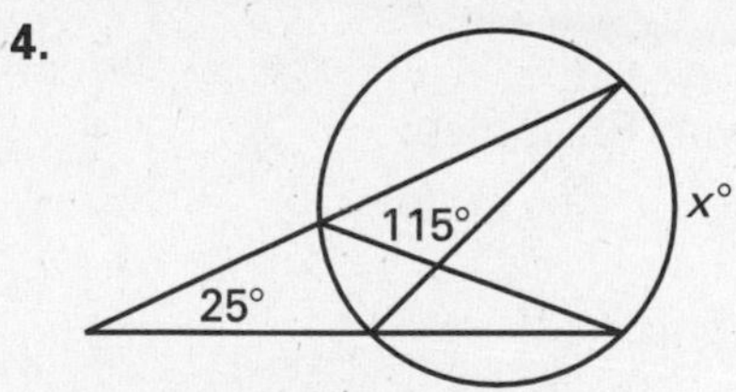

5.

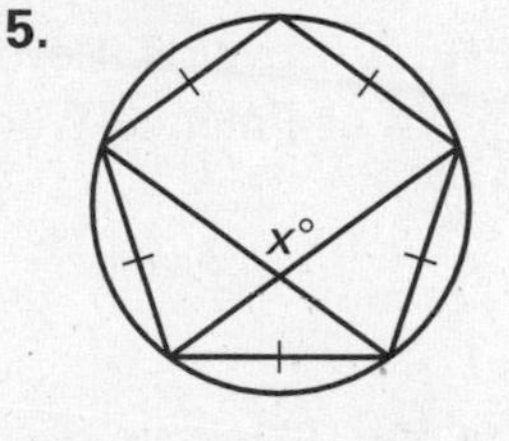

6.

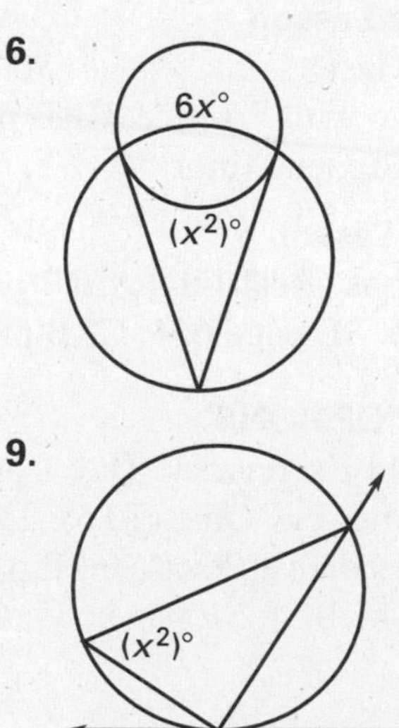

7.

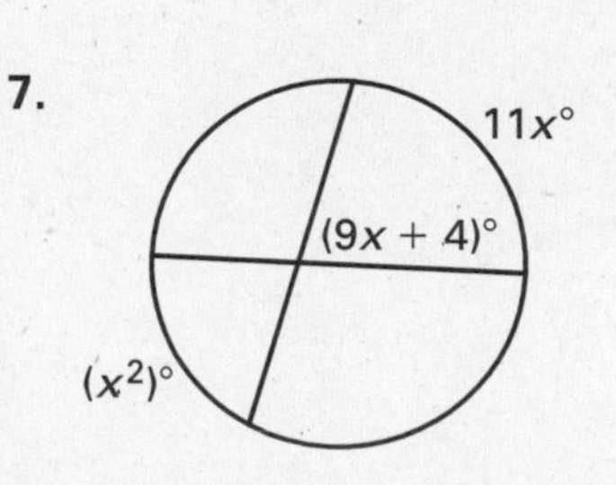

8.

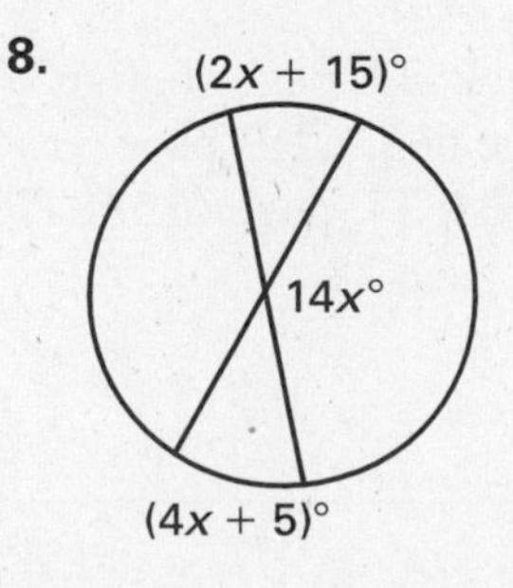

9.

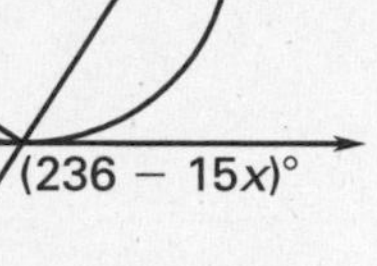

LESSON 10.5

TEACHER'S NAME  ______________________ CLASS ________ ROOM ________ DATE ________

Lesson Plan

2-day lesson (See *Pacing the Chapter,* TE pages 592C–592D) **For use with pages 628–635**

GOALS
1. **Find the lengths of segments of chords.**
2. **Find the lengths of segments of tangents and secants.**

State/Local Objectives ____

____

✓ Check the items you wish to use for this lesson.

STARTING OPTIONS

____ Homework Check: TE page 624: Answer Transparencies
____ Warm-Up or Daily Homework Quiz: TE pages 629 and 627, CRB page 66, or Transparencies

TEACHING OPTIONS

____ Motivating the Lesson: TE page 630
____ Lesson Opener (Activity): CRB page 67 or Transparencies
____ Technology Activity with Keystrokes: CRB pages 68–69
____ Examples: Day 1: 1–4, SE pages 629–631; Day 2: See the Extra Examples.
____ Extra Examples: Day 1 or Day 2: 1–4, TE pages 630–631 or Transp.
____ Technology Activity: SE page 628
____ Closure Question: TE page 631
____ Guided Practice: SE page 632 Day 1: Exs. 1–9; Day 2: See Checkpoint Exs. TE pages 630–631

APPLY/HOMEWORK

Homework Assignment

____ Basic Day 1: 10–30 even; Day 2: 11–31 odd, 33–35, 40–54 even; Quiz 2: 1–7
____ Average Day 1: 10–30 even; Day 2: 11–31 odd, 33–35, 40–54 even; Quiz 2: 1–7
____ Advanced Day 1: 10–32 even; Day 2: 11–31 odd, 33–39, 40–54 even; Quiz 2: 1–7

Reteaching the Lesson

____ Practice Masters: CRB pages 70–72 (Level A, Level B, Level C)
____ Reteaching with Practice: CRB pages 73–74 or Practice Workbook with Examples
____ Personal Student Tutor

Extending the Lesson

____ Applications (Real-Life): CRB page 76
____ Challenge: SE page 634; CRB page 77 or Internet

ASSESSMENT OPTIONS

____ Checkpoint Exercises: Day 1 or Day 2: TE pages 630–631 or Transp.
____ Daily Homework Quiz (10.5): TE page 635, CRB page 81, or Transparencies
____ Standardized Test Practice: SE page 634; TE page 635; STP Workbook; Transparencies
____ Quiz (10.4–10.5): SE page 635; CRB page 78

Notes ____

____

____

TEACHER'S NAME ______________ CLASS ______ ROOM ______ DATE ______

Lesson Plan for Block Scheduling

1-day lesson (See *Pacing the Chapter,* TE pages 592C–592D) **For use with pages 628–635**

1. **Find the lengths of segments of chords.**
2. **Find the lengths of segments of tangents and secants.**

State/Local Objectives ______________________________

CHAPTER PACING GUIDE	
Day	**Lesson**
1	10.1 (all)
2	10.2 (all)
3	10.3 (all)
4	10.4 (all); **10.5 (begin)**
5	**10.5 (end)**; 10.6 (all)
6	10.7 (all)
7	Review Ch. 10; Assess Ch. 10

✓ Check the items you wish to use for this lesson.

STARTING OPTIONS

____ Homework Check: TE page 624: Answer Transparencies
____ Warm-Up or Daily Homework Quiz: TE pages 629 and 627, CRB page 66, or Transparencies

TEACHING OPTIONS

____ Motivating the Lesson: TE page 630
____ Lesson Opener (Activity): CRB page 67 or Transparencies
____ Technology Activity with Keystrokes: CRB pages 68–69
____ Examples: Day 4: 1–4, SE pages 629–631; Day 5: See the Extra Examples.
____ Extra Examples: Day 4 or Day 5: 1–4, TE pages 630–631 or Transp.
____ Technology Activity: SE page 628
____ Closure Question: TE page 631
____ Guided Practice: SE page 632 Day 4: Exs. 1–9; Day 5: See Checkpoint Exs. TE pages 630–631

APPLY/HOMEWORK

Homework Assignment (See also the assignments for Lessons 10.4 and 10.6.)

____ Block Schedule: Day 4: 10–30 even; Day 5: 11–31 odd, 33–35, 40–54 even; Quiz 2: 1–7

Reteaching the Lesson

____ Practice Masters: CRB pages 70–72 (Level A, Level B, Level C)
____ Reteaching with Practice: CRB pages 73–74 or Practice Workbook with Examples
____ Personal Student Tutor

Extending the Lesson

____ Applications (Real-Life): CRB page 76
____ Challenge: SE page 634; CRB page 77 or Internet

ASSESSMENT OPTIONS

____ Checkpoint Exercises: Day 4 or Day 5: TE pages 630–631 or Transp.
____ Daily Homework Quiz (10.5): TE page 635, CRB page 81, or Transparencies
____ Standardized Test Practice: SE page 634; TE page 635; STP Workbook; Transparencies
____ Quiz (10.4–10.5): SE page 635; CRB page 78

Notes ______________________________

LESSON
10.5

NAME ______________________ DATE ________

Available as a transparency

WARM-UP EXERCISES

For use before Lesson 10.5, pages 628–635

Solve each equation.

1. $8x = 12 \cdot 4$

2. $8(10 + 8) = 6(y + 6)$

3. $10^2 = 6(2x + 8)$

4. $12^2 = x(x + 6)$

DAILY HOMEWORK QUIZ

For use after Lesson 10.4, pages 621–627

Find the value of *x*.

1. $m\widehat{HJK} = (20x - 4)°$

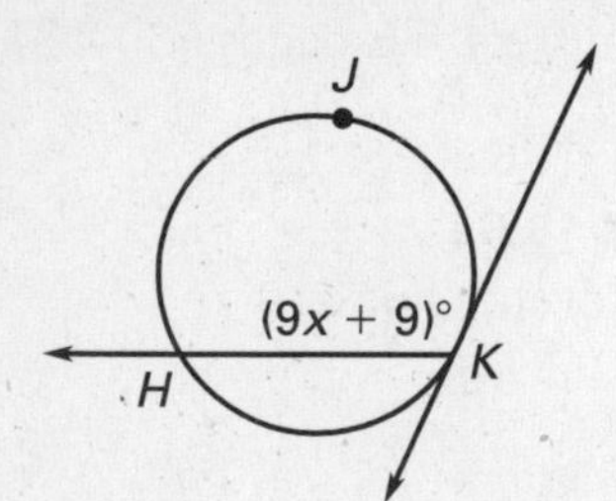

2.

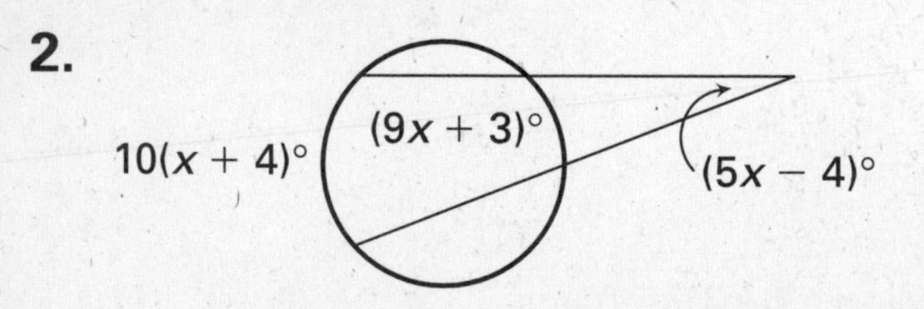

LESSON
10.5

NAME ______________________________ DATE __________

Available as a transparency

Lesson 10.5

Activity Lesson Opener

For use with pages 629–635

SET UP: Work in a group.
***You will need*: • compass • ruler • calculator**

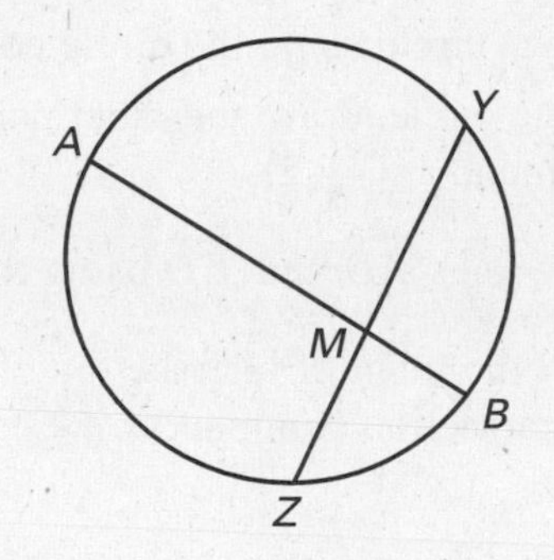

1. Each member of the group draws a large circle on a piece of paper. On your circle, draw any two chords $\overline{AB}$ and $\overline{YZ}$ that intersect inside the circle. Label M, the point of intersection of the chords. Measure carefully to find AM, MB, ZM, and MY to the nearest millimeter. Make any predictions about these lengths.

2. As a group, complete the table below by using measurements from your circles. Use a calculator to find the products.

Student's Name	*AM*	*MB*	*ZM*	*MY*	*AM · MB*	*ZM · MY*

3. Analyze your table. What do you notice about the two products for each circle? Compare your ideas and results with other groups in the class.

4. Make a conjecture about the lengths of the segments that are formed when two chords intersect inside a circle. Draw a diagram to support your conjecture.

LESSON 10.5

NAME ______________________ DATE __________

Technology Activity Keystrokes

For use with page 628

TI-92

Construct

1. Draw a circle using the circle command (F3 1).
2. On the circle, draw and label points A, B, C, and D. Use the point on the object command (F2 2).
3. Draw lines $\overleftrightarrow{AB}$ and $\overleftrightarrow{CD}$ using the line command (F2 4).
4. Draw the point of intersection of $\overleftrightarrow{AB}$ and $\overleftrightarrow{CD}$ and label the point E. Use the point of intersection command (F2 3).

Investigate

1. Drag points A, B, C, and D.
   F1 1 (Place cursor on point.) ENTER (Use the drag key [drag key] and the cursor pad to drag the point.)
2. Draw $\overline{EA}$, $\overline{EB}$, $\overline{EC}$, and $\overline{ED}$ using the segment command (F2 5). Hide $\overleftrightarrow{AB}$ and $\overleftrightarrow{CD}$ (F7 1).
3. Measure $\overline{EA}$, $\overline{EB}$, $\overline{EC}$, and $\overline{ED}$.
   F6 1 (Place cursor on segment EA.) ENTER
   Repeat this process for the other three segments.
   Calculate $EA \cdot EC$ and $EB \cdot ED$.
   F6 6 (Use cursor to highlight the length of segment $\overline{EA}$.) ENTER *
   (Highlight the length of $\overline{EC}$.) ENTER ENTER (The result will appear on the screen.) F6 6 (Use cursor to highlight the length of segment $\overline{EB}$.) ENTER *
   (Highlight the length of $\overline{ED}$.) ENTER ENTER (The result will appear on the screen.)
4. See Investigate Step 1.

NAME ____ DATE ____________

Reteaching with Practice

For use with pages 629–635

GOAL **Find the lengths of segments of chords, tangents, and secants**

VOCABULARY

In the figure shown, $\overline{PS}$ is a **tangent segment** because it is tangent to the circle at an endpoint. $\overline{PR}$ is a **secant segment** because one of the two intersection points with the circle is an endpoint. $\overline{PQ}$ is the **external segment** of $\overline{PR}$.

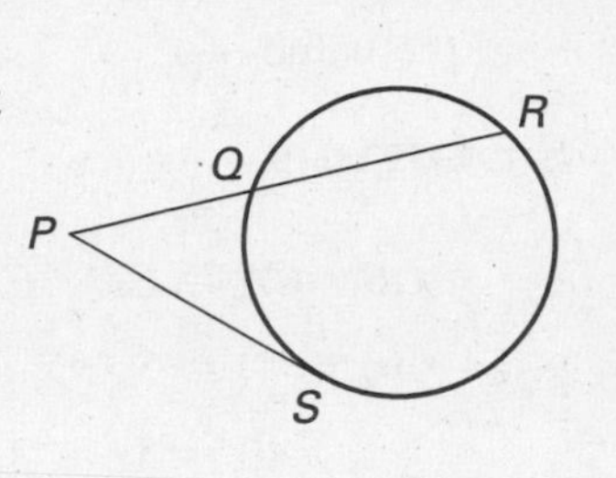

Theorem 10.16
If two secant segments share the same endpoint outside a circle, then the product of the length of one secant segment and the length of its external segment equals the product of the length of the other secant segment and the length of its external segment.

Theorem 10.17
If a secant segment and a tangent segment share an endpoint outside a circle, then the product of the length of the secant segment and the length of its external segment equals the square of the length of the tangent segment.

EXAMPLE 1 *Finding Segment Lengths Using Theorem 10.15*

Find the value of x.

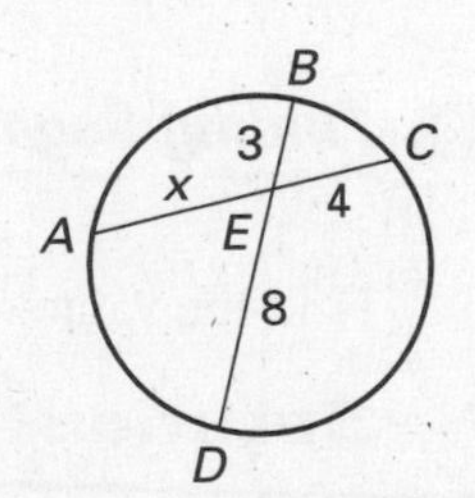

SOLUTION

Because $\overline{AC}$ and $\overline{BD}$ are chords which intersect in the interior of the circle, Theorem 10.15 applies.

$EC \cdot EA = EB \cdot ED$	Use Theorem 10.15.
$4 \cdot x = 3 \cdot 8$	Substitute.
$4x = 24$	Simplify.
$x = 6$	Divide each side by 4.

Exercises for Example 1

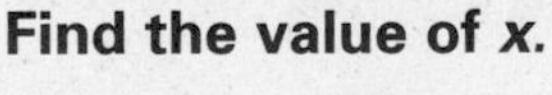

Find the value of x.

1.

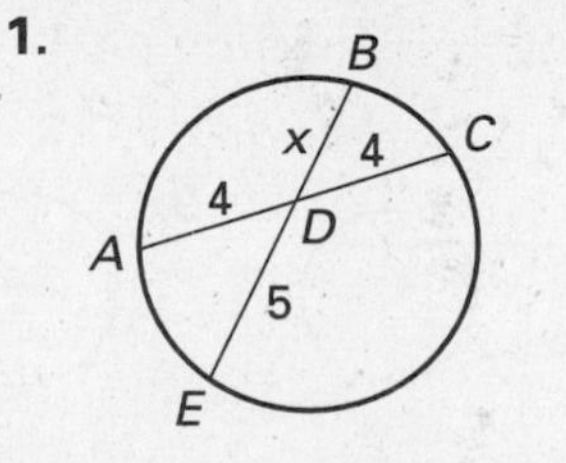

2.

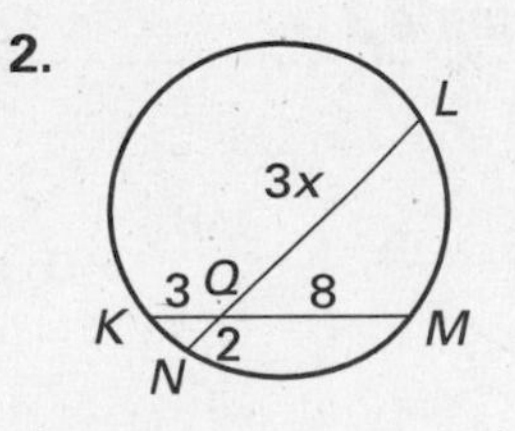

3.

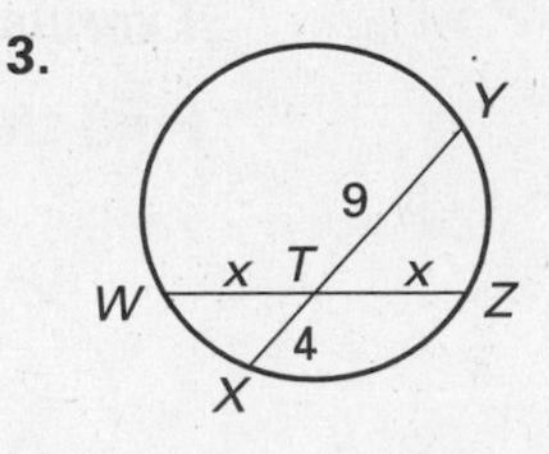

NAME ______________________________ DATE ____________

Reteaching with Practice

For use with pages 629–635

EXAMPLE 2 Finding Segment Lengths Using Theorem 10.16

Find the value of x.

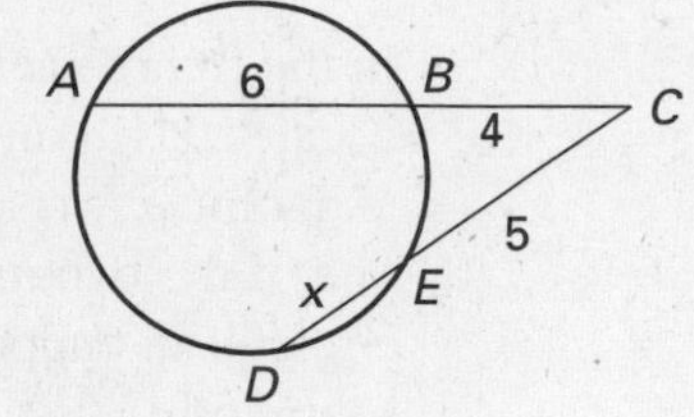

SOLUTION

$CB \cdot CA = CE \cdot CD$	Use Theorem 10.16.
$4 \cdot (6 + 4) = 5 \cdot (x + 5)$	Substitute.
$40 = 5x + 25$	Simplify.
$15 = 5x$	Subtract 25 from each side.
$x = 3$	Divide each side by 5.

Exercises for Example 2

Find the value of x.

4. **5.** **6.**

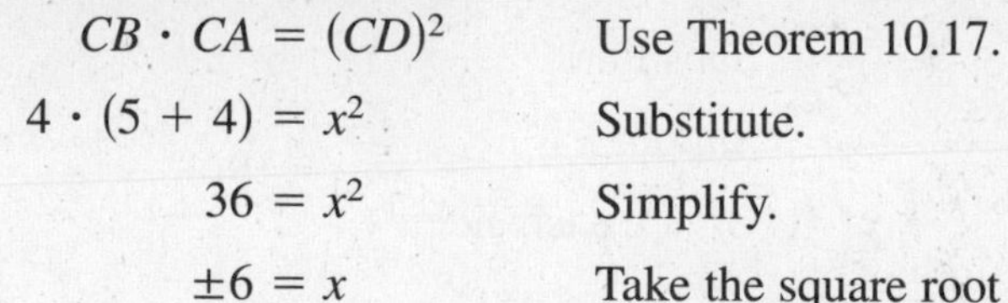

1
2x
1
4

EXAMPLE 3 Finding Segment Lengths Using Theorem 10.17

Find the value of x.

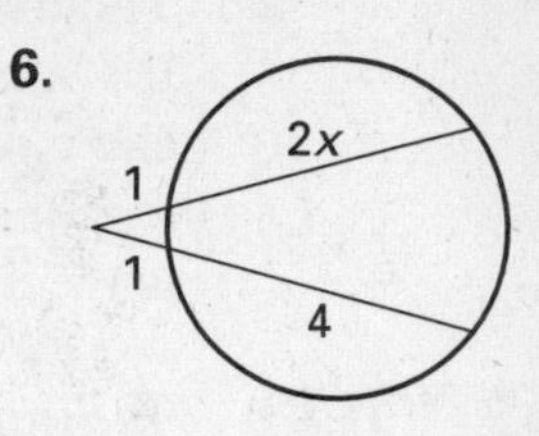

SOLUTION

$CB \cdot CA = (CD)^2$	Use Theorem 10.17.
$4 \cdot (5 + 4) = x^2$	Substitute.
$36 = x^2$	Simplify.
$\pm 6 = x$	Take the square root of each side.

Use the positive solution, because lengths cannot be negative. So, $x = 6$.

Exercises for Example 3

Find the value of x.

7. **8.** **9.**

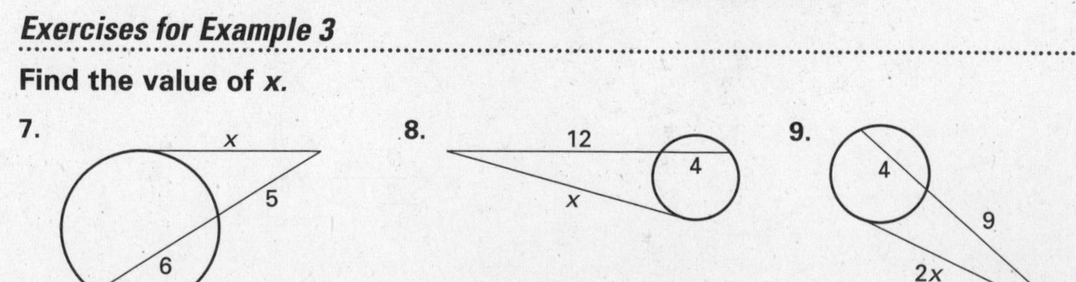

NAME ______________________________ DATE ____________

Quick Catch-Up for Absent Students

For use with pages 628–635

The items checked below were covered in class on (date missed) ____________

Activity 10.5: Investigating Segment Lengths (p. 628)

____ **Goal:** Use geometry software to explore the lengths of segments in a circle.

Lesson 10.5: Segment Lengths in Circles

____ **Goal 1:** Find the lengths of segments of chords. (p. 629)

Material Covered:

____ Example 1: Finding Segment Lengths

____ **Goal 2:** Find the lengths of segments of tangents and secants. (pp. 630–631)

Material Covered:

____ Example 2: Finding Segment Lengths

____ Example 3: Estimating the Radius of a Circle

____ Example 4: Finding Segment Lengths

Vocabulary:

tangent segment, p. 630
secant segment, p. 630
external segment, p. 630

____ Other (specify) ______________________________

____

Homework and Additional Learning Support

____ Textbook (specify) pp. 632–635

____

____ *Reteaching with Practice* worksheet (specify exercises)____________

____ *Personal Student Tutor* for Lesson 10.5

NAME ______________________ DATE ________

Real-Life Application: When Will I Ever Use This?

For use with pages 629–635

Saturn

The sixth planet from the sun and the second largest planet in the solar system is Saturn. Christiaan Huygens first discovered rings circling around Saturn in 1659. Two main rings (A and B) and one faint ring (C) can be seen from Earth. Pictures taken from Voyager I show four additional faint rings. Although these rings look continuous from Earth, the rings are actually made of countless small particles each in its own independent orbit. These ring particles appear to be composed primarily of ice. Some may be made of rocky particles with icy coverings. The origin of the rings of Saturn is unknown. Until very recently, Saturn was thought to be the only planet with rings circling around the planet. In 1977, faint rings were discovered around Uranus, Jupiter, and Neptune.

Saturn has 18 named satellites (or moons). Some of these satellites play a vital role in keeping the rings in place. One of these satellites, Prometheus, is the third of Saturn's known satellites and is located in the F-ring. Despite what is known about Saturn, its rings, and surrounding satellites, the whole system is very complex and is poorly understood.

On October 15, 1997, NASA, along with the European Space Agency, the Italian Space Agency, and several separate European partners launched the Cassini mission to explore Saturn, its system of rings, and the many satellites. The Cassini mission is set to arrive in Saturn's orbit on July 1, 2004.

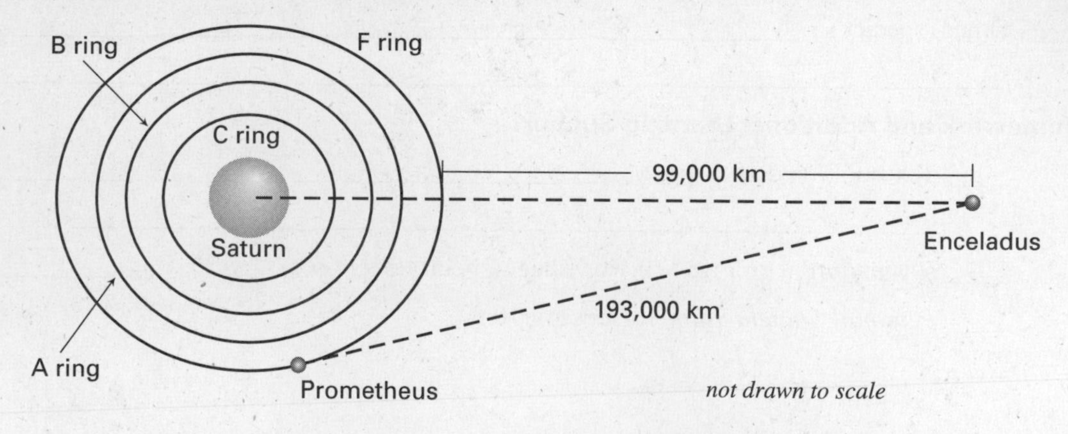

Use the diagram above to answer the following questions.

1. Describe the type of line segment between the two satellites Enceladus and Prometheus.
2. Estimate the radius from the center of Saturn to its F-Ring.
3. Find the distance between the center of Saturn and Enceladus.

NAME ______________________ DATE __________

Challenge: Skills and Applications

For use with pages 629–635

1. Refer to the diagram. Write a paragraph proof.

   Given: $\overleftrightarrow{OT}$ is tangent to both circles at T.

   Prove: $OP \cdot OQ = OR \cdot OS$

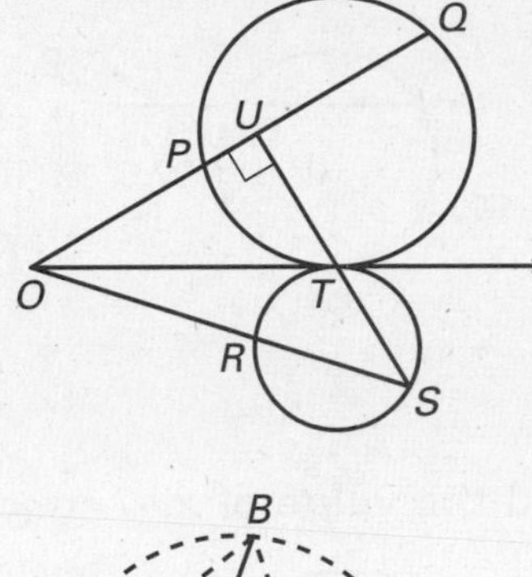

2. In this exercise, you will prove the converse of Theorem 10.15. Given line segments $\overline{AB}$ and $\overline{CD}$ intersecting at E, such that $EA \cdot EB = EC \cdot ED$, complete the following steps to show that there exists a circle containing A, B, C, and D.

   a. Draw $\overline{AC}$, $\overline{CB}$, $\overline{BD}$, and $\overline{DA}$. Show that $\triangle CEA \sim \triangle BED$ and $\triangle DEA \sim \triangle BEC$.

   b. Show that $m\angle ACB + m\angle ADB = 180°$.

   c. Use a theorem about quadrilaterals to show that there exists a circle containing A, B, C, and D.

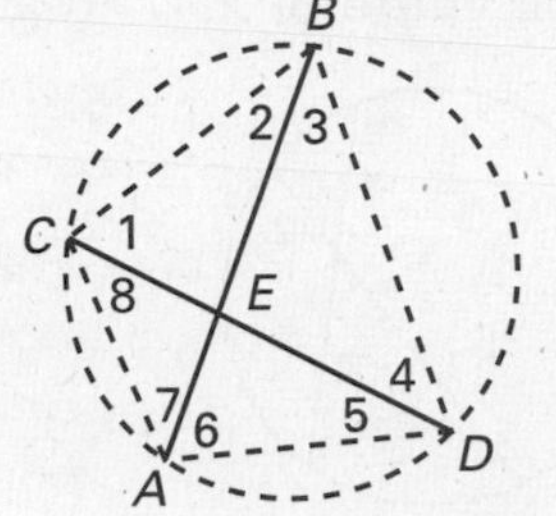

3. a. In the diagram at the right, $\overline{PR}$ is a perpendicular bisector of $\overline{SQ}$. Find $m\angle PQR$.

   b. Use a Geometric Mean Theorem to write a proportion involving $\triangle PQR$.

   c. Show how the theorem involving segments of chords could have been used (instead of the Geometric Mean Theorem) to obtain the result you found in part (b).

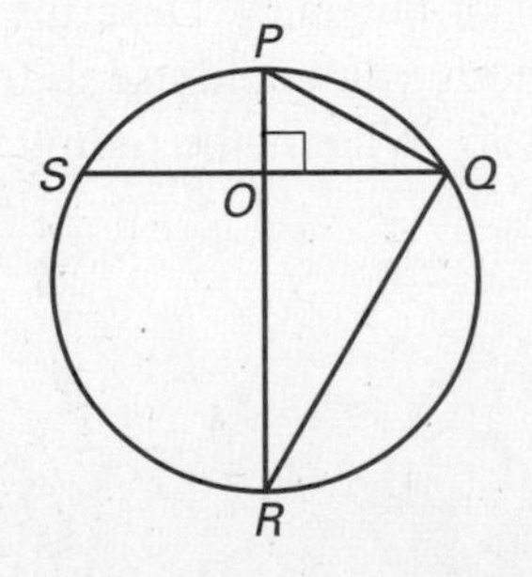

In Exercises 4–6, find the possible values of *x*.

4.

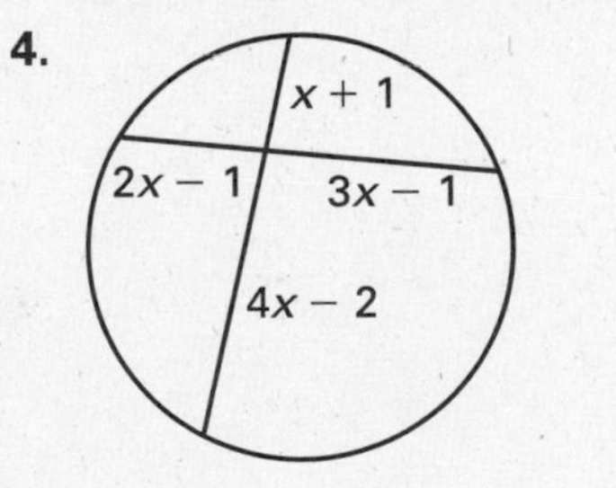

5.

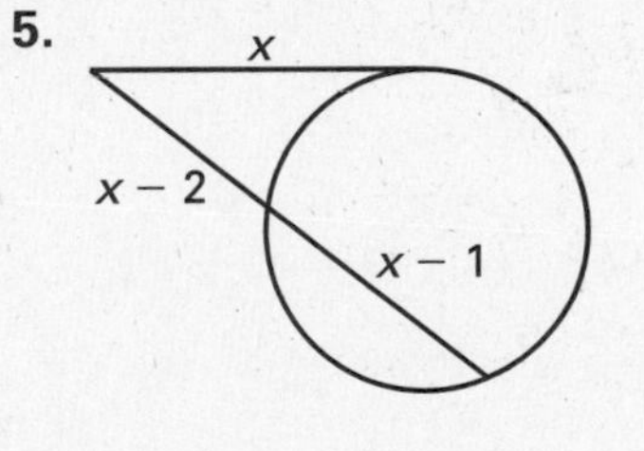

6.

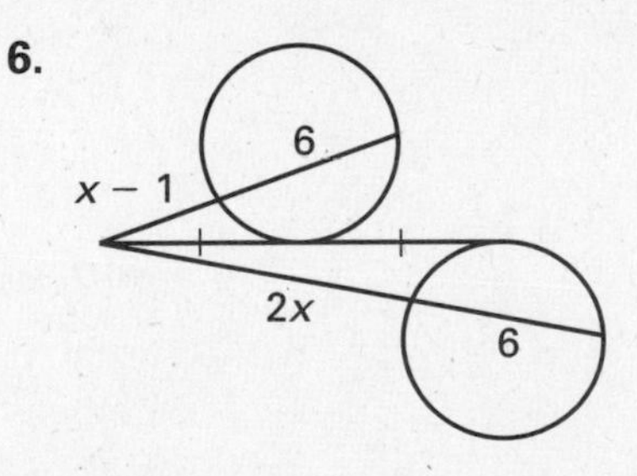

LESSON 10.5

NAME ______________________ DATE __________

Quiz 2

For use after Lessons 10.4 and 10.5

Find the value of *x*. *(Lesson 10.4)*

1. **2.** **3.**

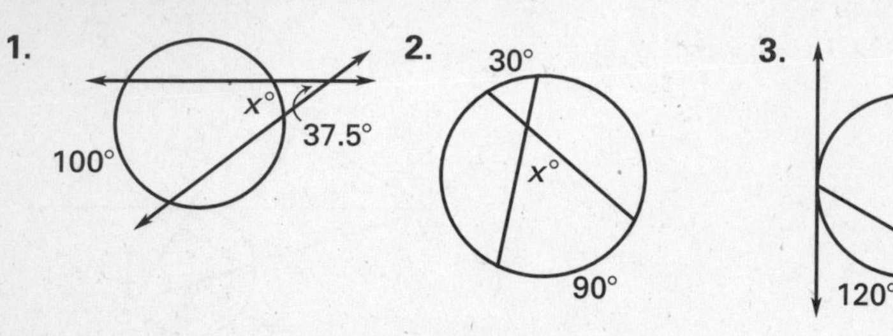

Find the value of *x*. *(Lesson 10.5)*

4. **5.**

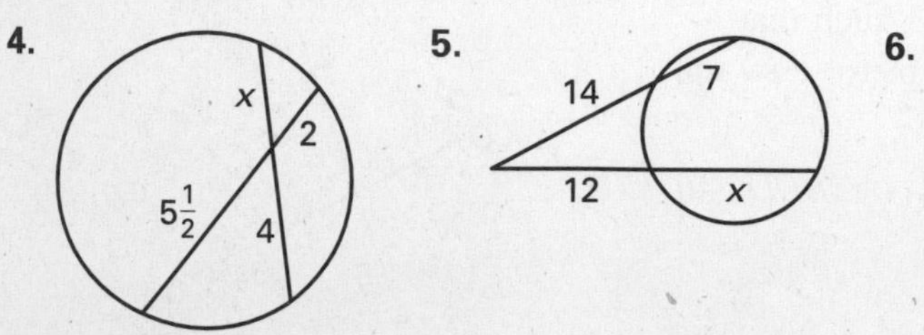

6.

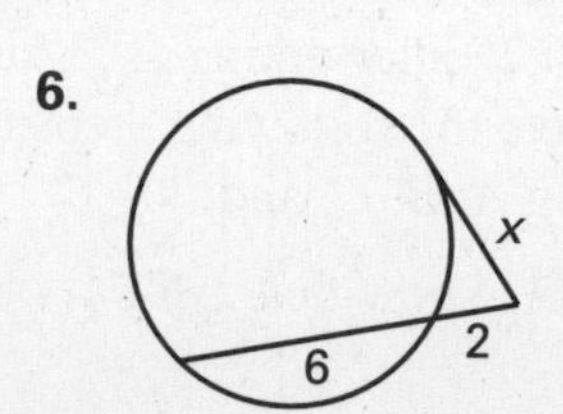

Answers

1. ______
2. ______
3. ______
4. ______
5. ______
6. ______
7. ______

7. ***Missile Silo*** You are standing 18 feet from the circular wall of a silo and 36 feet from a point of tangency. Describe two different methods you could use to find the radius of the silo. What is the radius? *(Lesson 10.5)*

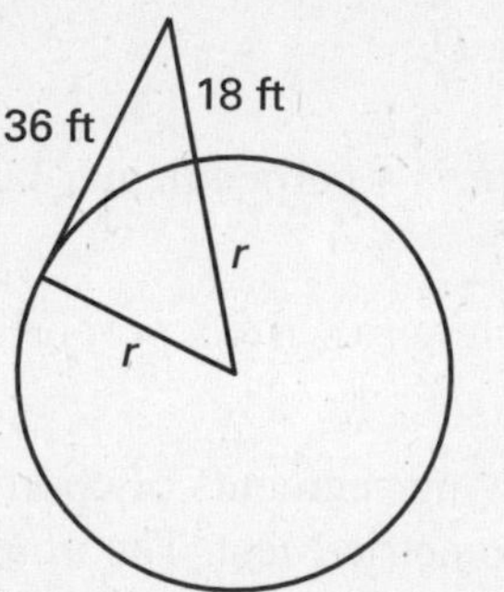

LESSON
10.6

TEACHER'S NAME ______________ CLASS ______ ROOM ______ DATE ______

Lesson Plan

1-day lesson (See *Pacing the Chapter,* TE pages 592C–592D) **For use with pages 636–640**

1. Write the equation of a circle.
2. Use the equation of a circle and its graph to solve problems.

State/Local Objectives ______________________________

____

Lesson 10.6

✓ Check the items you wish to use for this lesson.

STARTING OPTIONS

____ Homework Check: TE page 632: Answer Transparencies
____ Warm-Up or Daily Homework Quiz: TE pages 636 and 635, CRB page 81, or Transparencies

TEACHING OPTIONS

____ Lesson Opener (Visual Approach): CRB page 82 or Transparencies
____ Technology Activity with Keystrokes: CRB pages 83–85
____ Examples 1–4: SE pages 636–637
____ Extra Examples: TE page 637 or Transparencies
____ Closure Question: TE page 637
____ Guided Practice Exercises: SE page 638

APPLY/HOMEWORK

Homework Assignment

____ Basic 8–40 even, 47, 49, 50, 54–62 even
____ Average 8–40 even, 41, 42, 47–49, 50, 54–62 even
____ Advanced 8–40 even, 41–49, 50–53, 54–62 even

Reteaching the Lesson

____ Practice Masters: CRB pages 86–88 (Level A, Level B, Level C)
____ Reteaching with Practice: CRB pages 89–90 or Practice Workbook with Examples
____ Personal Student Tutor

Extending the Lesson

____ Cooperative Learning Activity: CRB page 92
____ Applications (Interdisciplinary): CRB page 93
____ Challenge: SE page 640; CRB page 94 or Internet

ASSESSMENT OPTIONS

____ Checkpoint Exercises: TE page 637 or Transparencies
____ Daily Homework Quiz (10.6): TE page 640, CRB page 97, or Transparencies
____ Standardized Test Practice: SE page 640; TE page 640; STP Workbook; Transparencies

Notes ______________________________

____

____

LESSON 10.6

TEACHER'S NAME ______________________ CLASS _______ ROOM _______ DATE _______

Lesson Plan for Block Scheduling

Half-day lesson (See *Pacing the Chapter,* TE pages 592C–592D) **For use with pages 636–640**

1. Write the equation of a circle.
2. Use the equation of a circle and its graph to solve problems.

State/Local Objectives ______________________

CHAPTER PACING GUIDE	
Day	**Lesson**
1	10.1 (all)
2	10.2 (all)
3	10.3 (all)
4	10.4 (all); 10.5 (begin)
5	10.5 (end); **10.6 (all)**
6	10.7 (all)
7	Review Ch. 10; Assess Ch. 10

✓ Check the items you wish to use for this lesson.

STARTING OPTIONS

____ Homework Check: TE page 632: Answer Transparencies
____ Warm-Up or Daily Homework Quiz: TE pages 636 and 635, CRB page 81, or Transparencies

TEACHING OPTIONS

____ Lesson Opener (Visual Approach): CRB page 82 or Transparencies
____ Technology Activity with Keystrokes: CRB pages 83–85
____ Examples 1–4: SE pages 636–637
____ Extra Examples: TE page 637 or Transparencies
____ Closure Question: TE page 637
____ Guided Practice Exercises: SE page 638

APPLY/HOMEWORK

Homework Assignment (See also the assignment for Lesson 10.5.)

____ Block Schedule: 8–40 even, 41, 42, 47–49, 50, 54–62 even

Reteaching the Lesson

____ Practice Masters: CRB pages 86–88 (Level A, Level B, Level C)
____ Reteaching with Practice: CRB pages 89–90 or Practice Workbook with Examples
____ Personal Student Tutor

Extending the Lesson

____ Cooperative Learning Activity: CRB page 92
____ Applications (Interdisciplinary): CRB page 93
____ Challenge: SE page 640; CRB page 94 or Internet

ASSESSMENT OPTIONS

____ Checkpoint Exercises: TE page 637 or Transparencies
____ Daily Homework Quiz (10.6): TE page 640, CRB page 97, or Transparencies
____ Standardized Test Practice: SE page 640; TE page 640; STP Workbook; Transparencies

Notes ______________________

LESSON **10.6**

NAME ______________________ DATE __________

Available as a transparency

WARM-UP EXERCISES

For use before Lesson 10.6, pages 636–640

Find the distance between each pair of points. Leave answers in simplest radical form.

1. $(0, 4), (0, -8)$
2. $(1, 5), (4, 1)$
3. $(-1, 2), (4, 1)$
4. $(0, 0), (4, 4)$
5. $(-3, -2), (2, 0)$

Lesson 10.6

DAILY HOMEWORK QUIZ

For use after Lesson 10.5, pages 628–635

Find the values of *x* and *y*.

1.

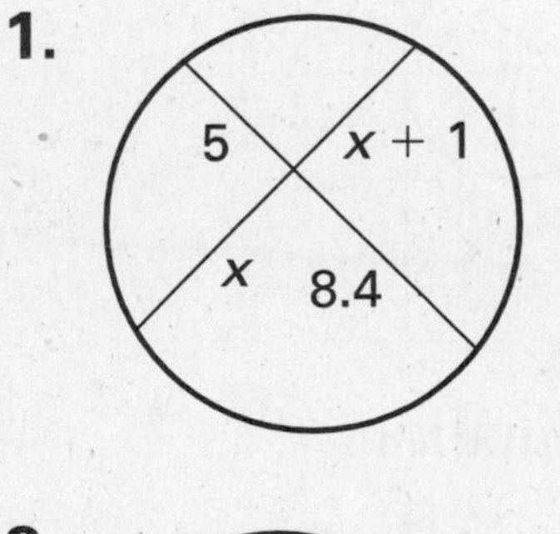

2.

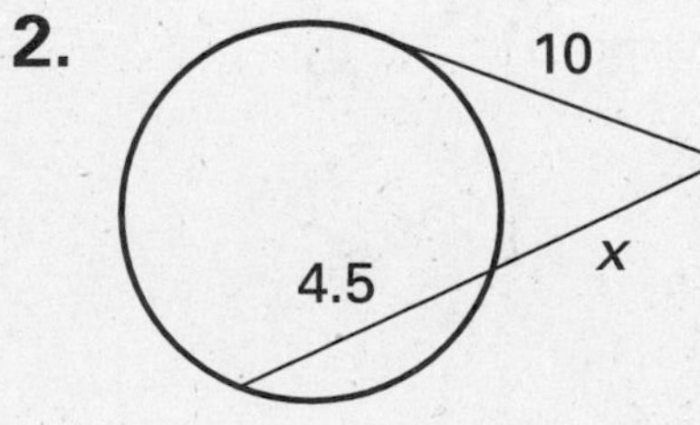

LESSON 10.6

NAME ______________________ DATE ________

Available as a transparency

Visual Approach Lesson Opener

For use with pages 636–640

Use the graphs and equations of 3 circles with center (0, 0).

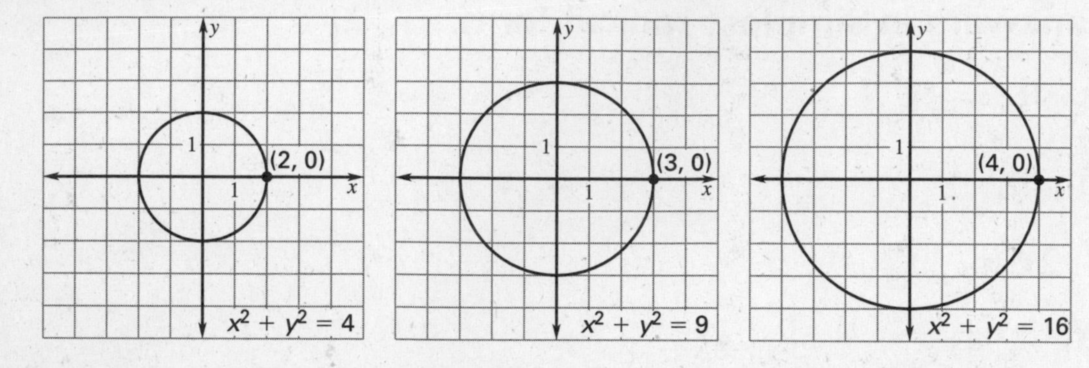

1. Graph the next circle in the pattern and write its equation.

2. Graph the circle that has equation $x^2 + y^2 = 81$.

Use the graphs and equations of 3 circles with different centers.

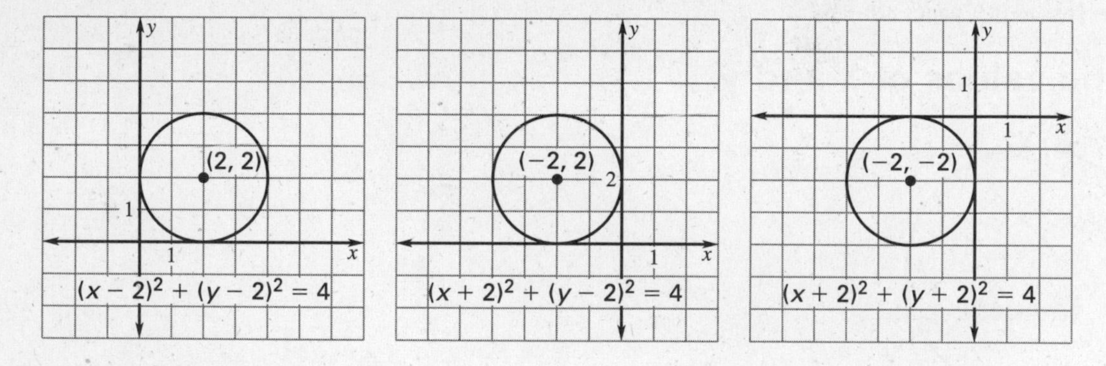

3. Graph the next circle in the pattern and write its equation.

4. Graph the circle that has equation $(x + 5)^2 + (y - 3)^2 = 9$.

NAME ______________________________ DATE __________

Technology Activity

For use with pages 636–640

GOAL **To analyze the relationship between a circle and its equation**

If you are given a circle in the coordinate plane, what is the equation of the circle? If you are given an equation of a circle, where is the circle located in the coordinate plane? In this activity you will determine the relationship between a circle and its equation.

Activity

1. Turn on the axes and grid.
2. Construct a circle with center C and a point on the circle labeled D.
3. Use the features of the geometry software to find the equation of the circle.
4. Drag the circle and observe the results.
5. Drag point D and observe the results.

Exercises

1. What is the relationship between a circle and its equation?

2. What is the center and radius of a circle with an equation of $(x + b)^2 + (y - c)^2 = d$?

3. Name two points on the circle given by the equation $(x - 3)^2 + (y - 5)^2 = 16$ that have an x-coordinate of 2.

4. Determine if the following points are on the circle, inside the circle, or outside the circle if the equation of the circle is $(x - 1)^2 + (y + 2)^2 = 9$.

a. $A(0, 1)$ **b.** $B(3, -3)$ **c.** $C(-2, -1)$ **d.** $D(-2, -2)$

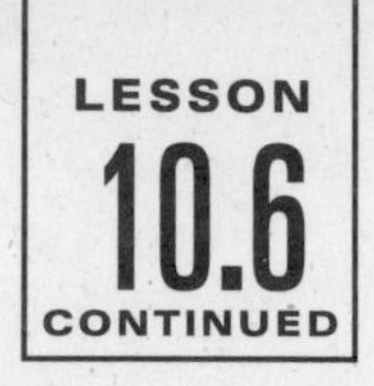

NAME ______________________________ DATE ____________

Technology Activity Keystrokes

For use with pages 636–640

TI-92

1. Turn on grid and axes.

   F8 9 (Set Coordinate Axes to RECTANGULAR and Grid to ON.) ENTER

2. Construct a circle with center *C* (at a grid point) and a point on the circle (at a grid point) labeled *D*.

   F3 1 (Move cursor to a grid point.) ENTER *C* (Move cursor until circle is desired size and at a grid point.) ENTER *D*

3. Find the equation of the circle.

   F6 5 (Place cursor on circle.) ENTER

4. Drag the circle and observe the results.

   F1 1 (Place cursor on *C*.) ENTER (Use the drag key and the cursor pad to drag the point.)

5. Drag *D* and observe the results.

   F1 1 (Place cursor on *D*.) ENTER (Use the drag key and the cursor pad to drag the point.)

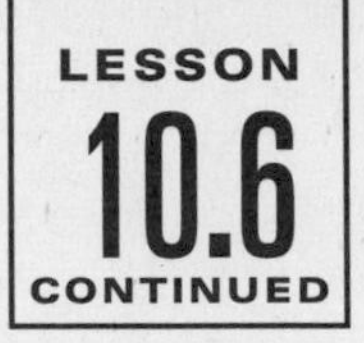

NAME ______________________________ DATE ____________

Technology Activity Keystrokes

For use with pages 636–640

SKETCHPAD

1. Turn on the grid and axes by selecting **Snap to Grid** from the **Graph** menu.
2. Use the compass tool to construct a circle with center C at a grid point with point D on the circle at a grid point.
3. Use the selection arrow tool to select the circle. Find the equation of the circle by choosing **Equation** from the **Measure** menu.
4. Use the translate selection arrow tool to select the circle and then drag it.
5. Use the translate selection arrow tool to select D and then drag it.

Lesson 10.6

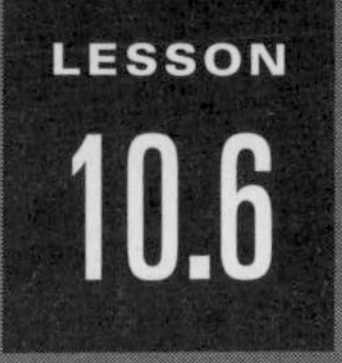

NAME ______________________ DATE ________

Practice A

For use with pages 636–640

Match the equation of a circle with its description.

1. $x^2 + y^2 = 4$
2. $x^2 + y^2 = 9$
3. $(x + 1)^2 + (y - 4)^2 = 16$
4. $(x + 2)^2 + (y + 3)^2 = 9$
5. $(x + 3)^2 + (y - 5)^2 = 16$
6. $(x - 2)^2 + (y - 5)^2 = 9$

A. center $(-1, 4)$, radius 4
B. center $(-2, -3)$, radius 3
C. center $(0, 0)$, radius 2
D. center $(2, 5)$, radius 3
E. center $(-3, 5)$, radius 4
F. center $(0, 0)$, radius 3

Give the center and radius of the circle.

7. $x^2 + y^2 = 25$
8. $x^2 + (y - 4)^2 = 9$
9. $(x - 5)^2 + y^2 = 16$
10. $(x + 1)^2 + (y - 1)^2 = 4$
11. $(x - 2)^2 + (y - 4)^2 = 16$
12. $(x + 4)^2 + (y - 2)^2 = 25$

Give the coordinates of the center, the radius, and the equation of the circle.

13.

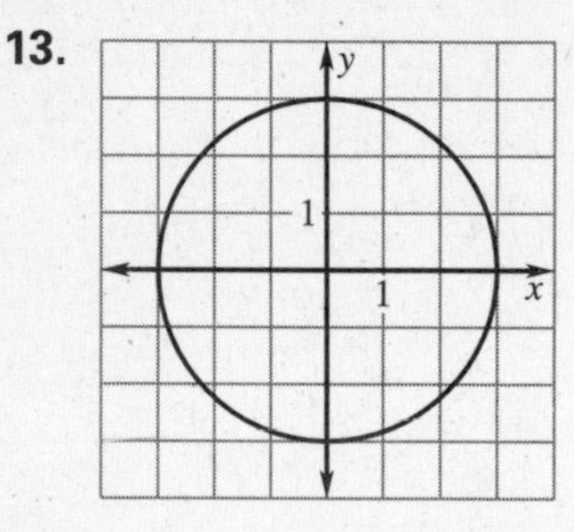

14.

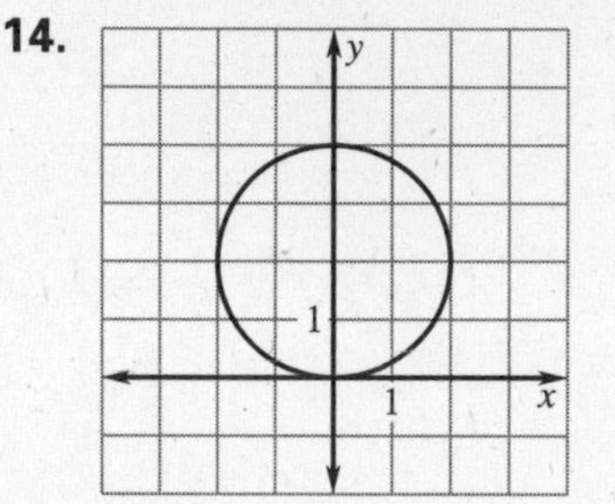

15.

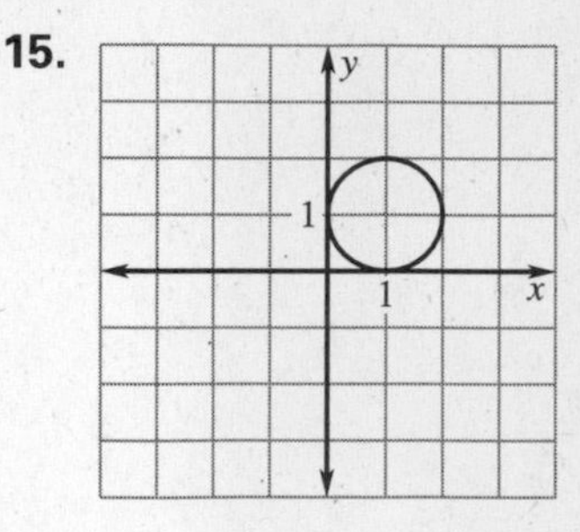

Write the standard equation of the circle with the given center and radius.

16. center $(0, 0)$, radius 2
17. center $(0, 1)$, radius 2
18. center $(2, 0)$, radius 3
19. center $(3, 3)$, radius 4

The equation of a circle is $(x - 2)^2 + (y - 2)^2 = 4$. Tell whether each point is *on* the circle, in the *interior* of the circle, or in the *exterior* of the circle.

20. $(1, 2)$
21. $(1, 4)$
22. $(2, 0)$
23. $(4, 2)$
24. $(4, 4)$
25. $(3, 2)$

NAME ______________________ DATE __________

Practice B

For use with pages 636–640

Match the equation of a circle with its description.

1. $(x + 2)^2 + (y + 3)^2 = 4$
2. $(x - 2)^2 + (y + 5)^2 = 9$
3. $(x + 3)^2 + (y - 5)^2 = 16$
4. $(x + 2)^2 + (y + 3)^2 = 9$
5. $(x - 3)^2 + (y + 5)^2 = 16$
6. $(x - 2)^2 + (y - 5)^2 = 9$

A. center $(-3, 5)$, radius 4
B. center $(-2, -3)$, radius 3
C. center $(-2, -3)$, radius 2
D. center $(2, 5)$, radius 3
E. center $(3, -5)$, radius 4
F. center $(2, -5)$, radius 3

Give the center and radius of the circle.

7. $(x - 4)^2 + (y + 2)^2 = 25$
8. $(x + 2)^2 + (y + 4)^2 = 9$
9. $(x - 5)^2 + (y - 3)^2 = 16$
10. $(x + 6)^2 + (y - 4)^2 = 4$
11. $(x - 5)^2 + (y - 6)^2 = 36$
12. $(x + 3)^2 + (y - 4)^2 = 16$

Give the coordinates of the center, the radius, and the equation of the circle.

13.
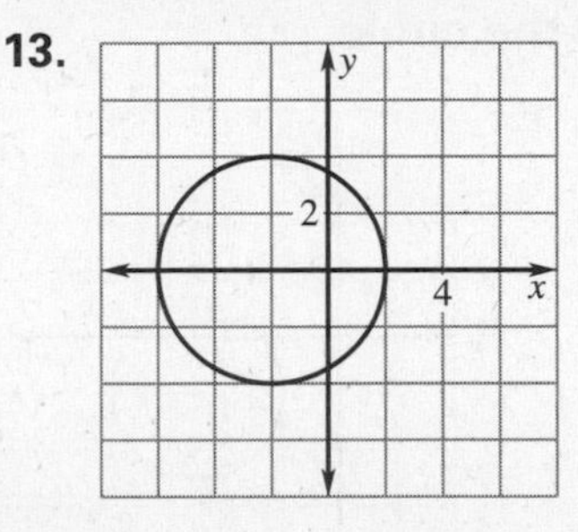

14.

15.
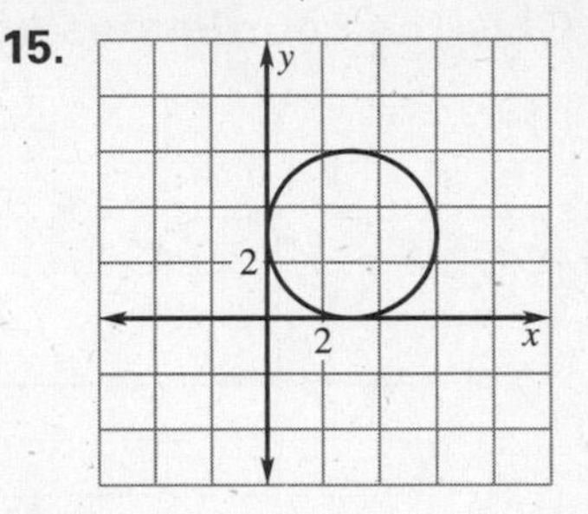

Write the standard equation of the circle with the given center and radius.

16. center $(0, 0)$, radius 1
17. center $(0, 4)$, radius 4
18. center $(-4, 2)$, radius 3
19. center $(-3, -5)$, radius 5

The equation of a circle is $(x - 4)^2 + (y - 2)^2 = 9$. Tell whether each point is *on* the circle, in the *interior* of the circle, or in the *exterior* of the circle.

20. $(5, 1)$
21. $(8, 2)$
22. $(1, 2)$
23. $(4, 5)$
24. $(0, 2)$
25. $(4, -2)$

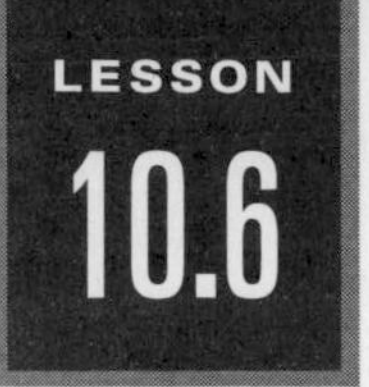

NAME ______________________________ DATE __________

Practice C

For use with pages 636–640

Match the equation of a circle with its description.

1. $(x + 2)^2 + (y - 3)^2 = 4$
2. $(x - 2)^2 + (y - 5)^2 = 4$
3. $(x + 3)^2 + (y - 5)^2 = 16$
4. $(x + 2)^2 + (y + 3)^2 = 4$
5. $(x + 3)^2 + (y + 5)^2 = 16$
6. $(x - 2)^2 + (y + 5)^2 = 4$

A. center $(-3, 5)$, radius 4
B. center $(-2, -3)$, radius 2
C. center $(-2, 3)$, radius 2
D. center $(2, -5)$, radius 2
E. center $(-3, -5)$, radius 4
F. center $(2, 5)$, radius 2

Give the center and radius of the circle.

7. $(x - 3)^2 + (y + 5)^2 = 36$
8. $(x + 4)^2 + (y + 2)^2 = 81$
9. $(x - 9)^2 + (y - 5)^2 = 40$
10. $(x + 1.5)^2 + (y - 3.8)^2 = 1.44$
11. $\left(x - \frac{1}{2}\right)^2 + \left(y - \frac{3}{4}\right)^2 = \frac{4}{9}$
12. $\left(x + \frac{3}{5}\right)^2 + \left(y - \frac{1}{10}\right)^2 = \frac{9}{25}$

Give the coordinates of the center, the radius, and the equation of the circle.

13.
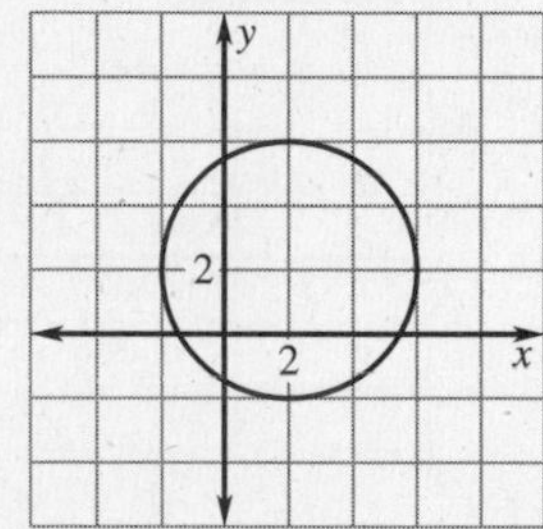

14.
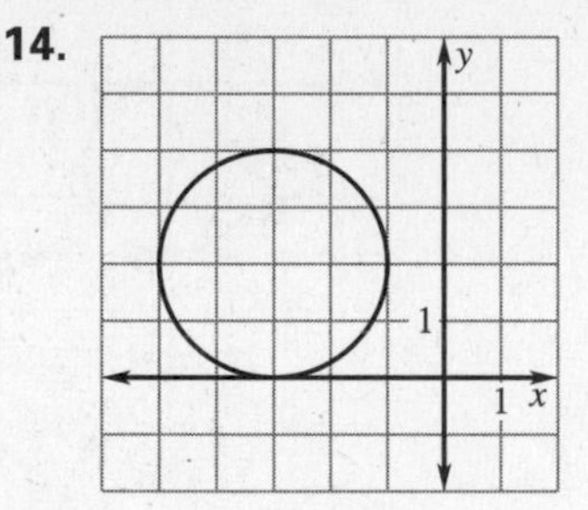

15.
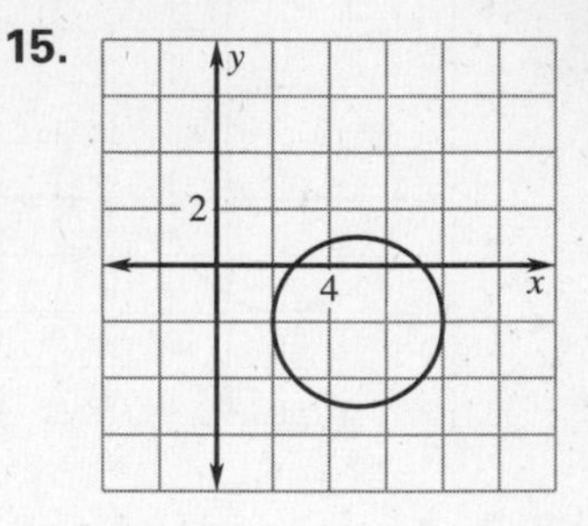

Write the standard equation of the circle with the given center and radius.

16. center $(0, 4)$, radius 5
17. center $(-3, 6)$, radius 7
18. center $(4.2, 2.6)$, radius 3.5
19. center $\left(\frac{7}{2}, \frac{5}{2}\right)$, radius 2

Graph the equation.

20. $(x - 3)^2 + (y + 4)^2 = 16$
21. $(x + 5)^2 + (y - 7)^2 = 25$

Graph the circle $(x - 4)^2 + (y + 2)^2 = 16$ and the line having the given equation. Determine whether the line is a tangent or a secant. Explain.

22. $y = x - 2$
23. $y = 2$
24. $y = -x + 6$

LESSON

10.6

NAME ______________________ DATE ____________

Reteaching with Practice

For use with pages 636–640

GOAL **Write the equation of a circle and use it and its graph to solve problems**

VOCABULARY

The **standard equation of a circle** with radius r and center (h, k) is $(x - h)^2 + (y - k)^2 = r^2$.

Lesson 10.6

EXAMPLE 1 *Writing a Standard Equation of a Circle*

a. Write the standard equation of the circle with center (2, 4) and radius 5.

b. The point $(-2, 4)$ is on a circle whose center is (0, 2). Write the standard equation of the circle.

SOLUTION

a. $(x - h)^2 + (y - k)^2 = r^2$ Standard equation of a circle

$(x - 2)^2 + (y - 4)^2 = 5^2$ Substitute.

$(x - 2)^2 + (y - 4)^2 = 25$ Simplify.

b. The radius is the distance from the point $(-2, 4)$ to the center (0, 2).

$r = \sqrt{(0 - (-2))^2 + (2 - 4)^2}$ Use the Distance Formula.

$r = \sqrt{2^2 + (-2)^2} = 2\sqrt{2}$ Simplify.

Substitute $(h, k) = (0, 2)$ and $r = 2\sqrt{2}$ into the standard equation of a circle.

$(x - h)^2 + (y - k)^2 = r^2$ Standard equation of a circle

$(x - 0)^2 + (y - 2)^2 = \left(2\sqrt{2}\right)^2$ Substitute.

$x^2 + (y - 2)^2 = 8$ Simplify.

Exercises for Example 1

Write the standard equation of the circle described.

1. center $(4, -1)$, radius 6

2. center $(-1, -5)$, radius 3.2

3. The center is $(-2, 3)$, a point on the circle is (2, 3).

EXAMPLE 2 *Graphing a Circle*

The equation of a circle is $(x - 1)^2 + (y + 3)^2 = 25$. Graph the circle.

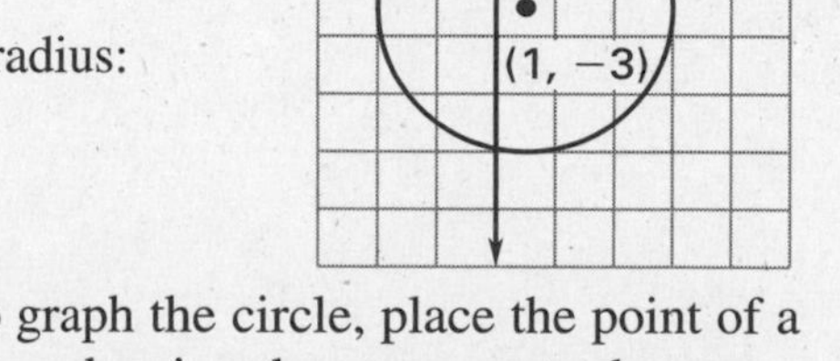

SOLUTION

Rewrite the equation to find the center and radius:

$(x - 1)^2 + (y + 3)^2 = 25$

$(x - 1)^2 + (y - (-3))^2 = 5^2$

The center is $(1, -3)$ and the radius is 5. To graph the circle, place the point of a compass at $(1, -3)$, set the radius at 5 units, and swing the compass to draw a full circle.

NAME ______________________ DATE ________

Reteaching with Practice

For use with pages 636–640

Exercises for Example 2

Graph the circle that has the given equation.

4. $(x - 2)^2 + (y - 7)^2 = 4$

5. $(x + 6)^2 + (y - 4)^2 = 9$

6. $(x + 3)^2 + y^2 = 16$

7. $x^2 + (y + 2)^2 = \frac{1}{2}$

EXAMPLE 3 Applying Graphs of Circles

A farmer's plot of land was struck by a meteorite which damaged a circular area of his farm. If the farmer's house is labeled as the origin of a coordinate plane, the area damaged by the meteorite can be expressed by the equation $(x - 6)^2 + (y - 7)^2 = 16$.

a. Graph the damaged area of the farm.

b. Items on the farm are located as follows: A silo is at (2, 4), a barn is at (4, 6), and a pigpen is at (8, 9). Which of these items were damaged by the meteorite?

SOLUTION

a. Rewrite the equation to find the center and radius:

$$(x - 6)^2 + (y - 7)^2 = 16$$

$$(x - 6)^2 + (y - 7)^2 = 4^2$$

The center is (6, 7) and the radius is 4.

b. The graph shows that the barn and the pigpen were damaged by the meteorite.

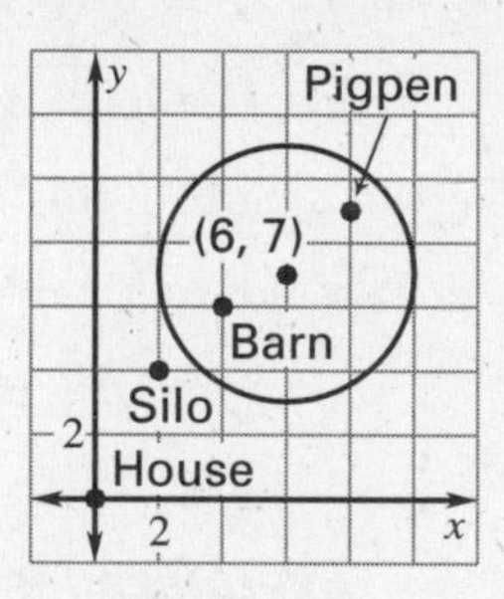

Exercises for Example 3

In Exercises 8–10, reconsider the situation from Example 3 above, assuming that the damage from the meteorite can be expressed by the equation $(x - 3)^2 + (y - 3)^2 = 9$. Did the meteorite damage the following items in this new situation?

8. The farmer's house

9. The silo

10. The pigpen

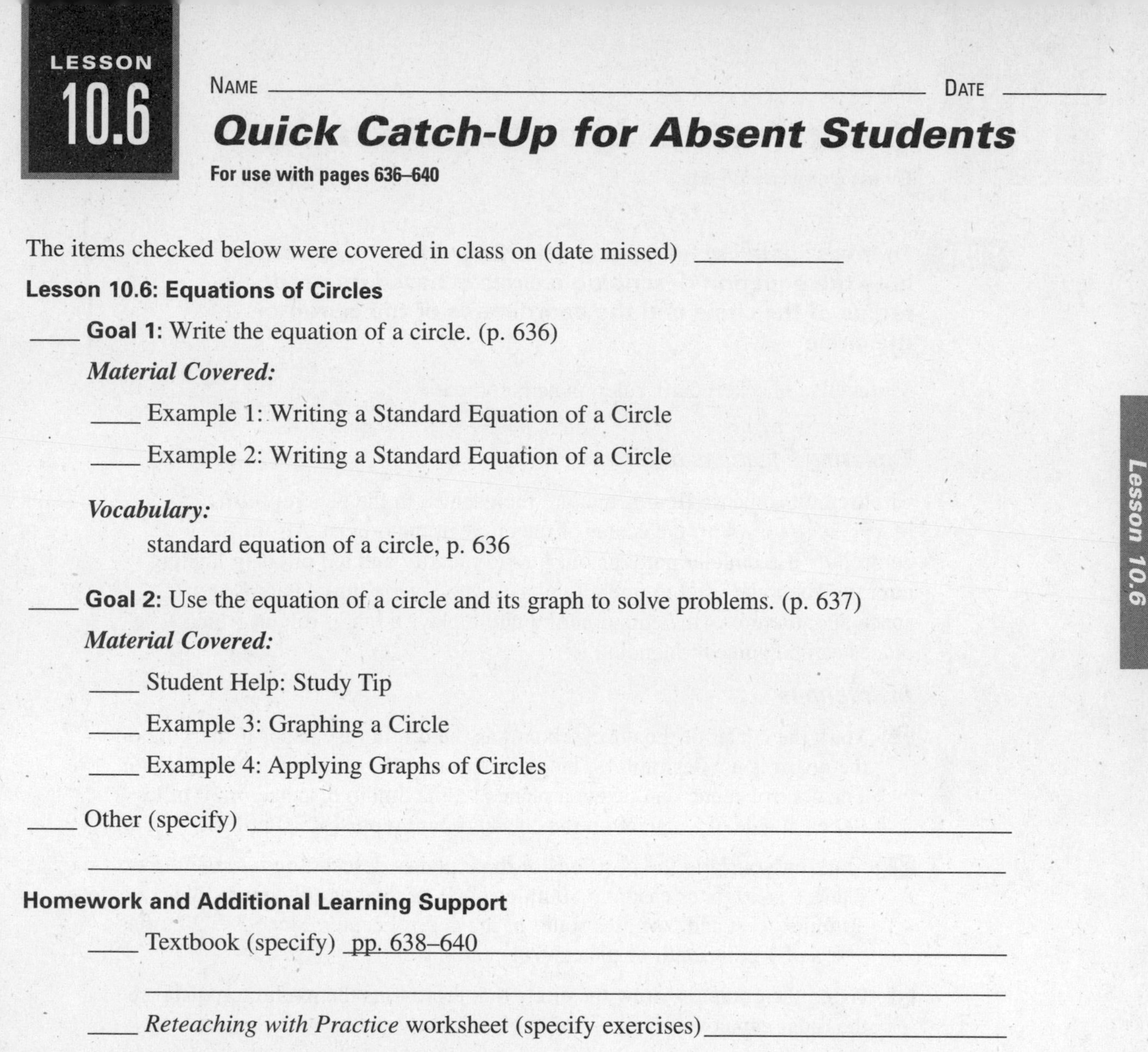

LESSON 10.6

NAME ______________________________ DATE __________

Quick Catch-Up for Absent Students

For use with pages 636–640

The items checked below were covered in class on (date missed) __________

Lesson 10.6: Equations of Circles

____ **Goal 1:** Write the equation of a circle. (p. 636)

Material Covered:

____ Example 1: Writing a Standard Equation of a Circle

____ Example 2: Writing a Standard Equation of a Circle

Vocabulary:

standard equation of a circle, p. 636

____ **Goal 2:** Use the equation of a circle and its graph to solve problems. (p. 637)

Material Covered:

____ Student Help: Study Tip

____ Example 3: Graphing a Circle

____ Example 4: Applying Graphs of Circles

____ Other (specify) ______________________________

____

Homework and Additional Learning Support

____ Textbook (specify) pp. 638–640

____

____ *Reteaching with Practice* worksheet (specify exercises) __________

____ *Personal Student Tutor* for Lesson 10.6

NAME ______________________ DATE ________

Cooperative Learning Activity

For use with pages 636–640

GOAL **To investigate circles on a checkerboard and to determine how the equation describing a circle is based upon the radius of the circle and the coordinates of the center of the circle**

Materials: checkerboard, ruler, pencil, compass

Exploring Equations of Circles

Circles are geometric figures that are represented in the general form of $x^2 + y^2 = r^2$ (with the center of the circle at the origin). Air traffic controllers use circular patterns of radar to identify and aid pilots in landing aircraft. The radar used in tracking planes can cover a limited amount of space and, therefore, the equation of a circle plays a major role in which planes can be guided to landing.

Instructions

1. Mark the center of the checkerboard as the origin (use paper if marking on the board is not desirable). This will represent the air traffic control center. The control center can detect a plane at a maximum distance of 30 miles (let each side of a square on the checkerboard represent 10 miles).
2. Mark on your grid the positions of three planes descending into the airport. Plane 1 is 20 miles east and 50 miles north of the control center. Plane 2 is 30 miles west and 20 miles south of the control center. Plane 3 is 20 miles east and 1 mile south of the control center.
3. Using the compass, draw the circle that represents the maximum distance the radar can cover.

Analyzing the Results

1. What is the equation of the circle that represents the maximum distance the radar can cover?
2. Which, if any, of the planes are within radar distance of the control center based upon the coordinate grid on the checkerboard?
3. What part of the circle represents the maximum distance that the radar can cover?

NAME ______________________ DATE __________

Real-Life Application: When Will I Ever Use This?

For use with pages 636–640

Circle Game

GAMES Suppose you are on the Spring Fair committee at your school and you want to design a game to be played at the fair. You can use your knowledge of circles to design the game.

A series of concentric circles is placed at the end of a ramp. A player rolls a ball down the ramp. At the end of the ramp, the ball will bounce into one of the circles. The inner circle is worth 12 points, the next circle is worth 9 points, the third circle is worth 3 points, and the outer circle is worth 1 point.

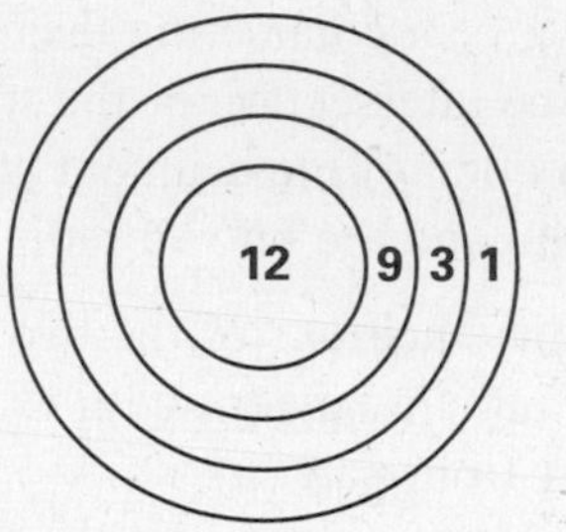

You assume that, on average:

- 5% of the balls will bounce into the 12-point circle.
- 15% of the balls will bounce into the 9-point circle.
- 30% of the balls will bounce into the 3-point circle.
- 50% of the balls will bounce into the 1-point circle.

In Exercises 1–4, use the information above and the following information.

Assume that your design for this game includes a coordinate plane with the circles plotted. The center of the circles is at (8, 4). The radius of each circle is given in the table.

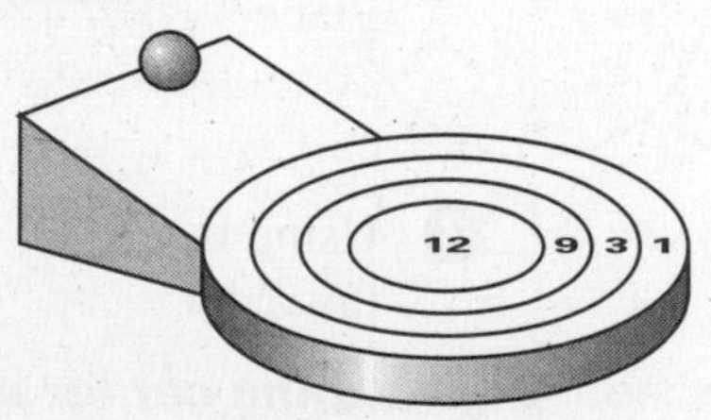

Circle	12-point circle	9-point circle	3-point circle	1-point circle
Radius	2 feet	3 feet	4 feet	5 feet

1. Write the equations of the circles.
2. Graph your equations from Exercise 1 in a coordinate plane. Each square unit is one square foot.
3. During a turn, a player rolls 4 balls down the ramp. Use your graph from Exercise 2 to find the total point value for a turn where the four balls land at the following locations: (7, 4), (10, 7), (10, 0), and (9, 6).
4. During the entire day of the Fair, 1200 people played the game, with each person rolling 4 balls down the ramp. On average, how many balls rolled into each of the circles described above?

NAME ______________________ DATE __________

Challenge: Skills and Applications

For use with pages 636–640

1. Circle C_1 has equation $(x + 2)^2 + (y + 4)^2 = 64$, and circle C_2 has equation $(x - h)^2 + (y - 1)^2 = 81$. The distance between the centers of the circles is 13.
   a. Find all possible values of h.
   b. If a segment connecting the centers of the circles is drawn, let A be the intersection of the segment with C_1, and let B be the intersection of the segment with C_2. Find AB. (*Hint:* Use a diagram and the known radii.)
   c. Find the equations of the two circles that have the same center as C_1 and are tangent with C_2. (*Hint:* Use your diagram and answer from part (b).)

2. Use the diagram to write a coordinate proof showing that the perpendicular bisector of a chord of a circle contains the center of the circle. (Let M be the midpoint of the chord.)

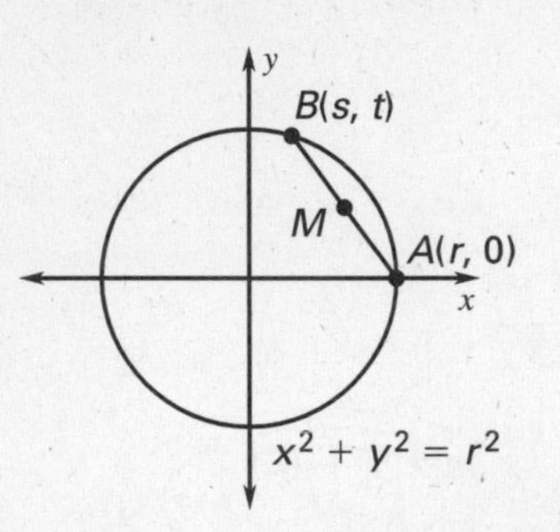

3. Use the diagram to write a coordinate proof showing that if one side of a triangle inscribed in a circle is a diameter of the circle, then the triangle is a right triangle.

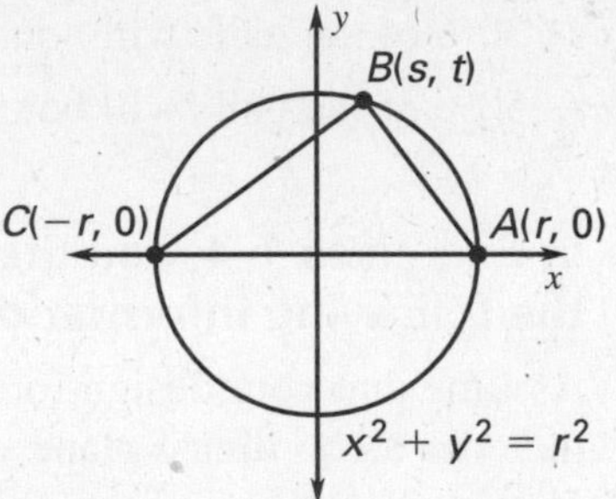

In Exercises 4–9, find the center and radius of the circle.

Example: $x^2 + y^2 - 6x + 4y - 3 = 0$

Complete the square in x and in y.

$$x^2 - 6x \qquad + y^2 + 4y \qquad = 3$$
$$(x^2 - 6x + 9) + (y^2 + 4y + 4) = 3 + 9 + 4$$
$$(x - 3)^2 + (y + 2)^2 = 16$$

The circle has center $(3, -2)$ and radius 4.

4. $x^2 + y^2 + 4x + 6y - 36 = 0$
5. $x^2 + y^2 - 10x + 8y - 23 = 0$
6. $x^2 + y^2 + 2x - 35 = 0$
7. $x^2 + y^2 + 6x - 8y = 0$
8. $x^2 + y^2 + 6x - 14y - 12 = 0$
9. $x^2 + y^2 - 8x - 4y + 18 = 0$

TEACHER'S NAME ______________________ CLASS ______ ROOM ______ DATE ______

Lesson Plan

2-day lesson (See *Pacing the Chapter,* TE pages 592C–592D) **For use with pages 641–648**

1. Draw the locus of points that satisfy a given condition.
2. Draw the locus of points that satisfy two or more conditions.

State/Local Objectives ____

____

✓ Check the items you wish to use for this lesson.

STARTING OPTIONS

____ Homework Check: TE page 638: Answer Transparencies
____ Warm-Up or Daily Homework Quiz: TE pages 642 and 640, CRB page 97, or Transparencies

TEACHING OPTIONS

____ Motivating the Lesson: TE page 643
____ Lesson Opener (Activity): CRB page 98 or Transparencies
____ Technology Activity with Keystrokes: CRB pages 99–101
____ Examples: Day 1: 1–4, SE pages 642–644; Day 2: See the Extra Examples.
____ Extra Examples: Day 1 or Day 2: 1–4, TE pages 643–644 or Transp.
____ Technology Activity: SE page 641
____ Closure Question: TE page 644
____ Guided Practice: SE page 645 Day 1: Exs. 1–8; Day 2: See Checkpoint Exs. TE pages 643–644

APPLY/HOMEWORK

Homework Assignment

____ Basic Day 1: 10–26 even, 28–30; Day 2: 9–27 odd, 33, 34, 36–45; Quiz 3: 1–8
____ Average Day 1: 10–26 even, 28–30; Day 2: 9–27 odd, 33, 34, 36–45; Quiz 3: 1–8
____ Advanced Day 1: 10–26 even, 28–30; Day 2: 9–27 odd, 32–45; Quiz 3: 1–8

Reteaching the Lesson

____ Practice Masters: CRB pages 102–104 (Level A, Level B, Level C)
____ Reteaching with Practice: CRB pages 105–106 or Practice Workbook with Examples
____ Personal Student Tutor

Extending the Lesson

____ Applications (Real-Life): CRB page 108
____ Math & History: SE page 648; CRB page 109; Internet
____ Challenge: SE page 647; CRB page 110 or Internet

ASSESSMENT OPTIONS

____ Checkpoint Exercises: Day 1 or Day 2: TE pages 643–644 or Transp.
____ Daily Homework Quiz (10.7): TE page 648, or Transparencies
____ Standardized Test Practice: SE page 647; TE page 648; STP Workbook; Transparencies
____ Quiz (10.6–10.7): SE page 648

Notes ____

____

____

TEACHER'S NAME ______________________ CLASS ______ ROOM ______ DATE ______

Lesson Plan for Block Scheduling

1-day lesson (See *Pacing the Chapter,* TE pages 592C–592D) **For use with pages 641–648**

GOALS
1. **Draw the locus of points that satisfy a given condition.**
2. **Draw the locus of points that satisfy two or more conditions.**

State/Local Objectives ____

CHAPTER PACING GUIDE	
Day	**Lesson**
1	10.1 (all)
2	10.2 (all)
3	10.3 (all)
4	10.4 (all); 10.5 (begin)
5	10.5 (end); 10.6 (all)
6	**10.7 (all)**
7	Review Ch. 10; Assess Ch. 10

✓ Check the items you wish to use for this lesson.

STARTING OPTIONS

____ Homework Check: TE page 638: Answer Transparencies
____ Warm-Up or Daily Homework Quiz: TE pages 642 and 640, CRB page 97, or Transparencies

TEACHING OPTIONS

____ Motivating the Lesson: TE page 643
____ Lesson Opener (Activity): CRB page 98 or Transparencies
____ Technology Activity with Keystrokes: CRB pages 99–101
____ Examples 1–4: SE pages 642–644
____ Extra Examples: TE pages 643–644 or Transparencies
____ Technology Activity: SE page 641
____ Closure Question: TE page 644
____ Guided Practice Exercises: SE page 645

APPLY/HOMEWORK

Homework Assignment

____ Block Schedule: 9–30, 33, 34, 36–45; Quiz 3: 1–8

Reteaching the Lesson

____ Practice Masters: CRB pages 102–104 (Level A, Level B, Level C)
____ Reteaching with Practice: CRB pages 105–106 or Practice Workbook with Examples
____ Personal Student Tutor

Extending the Lesson

____ Applications (Real-Life): CRB page 108
____ Math & History: SE page 648; CRB page 109; Internet
____ Challenge: SE page 647; CRB page 110 or Internet

ASSESSMENT OPTIONS

____ Checkpoint Exercises: TE pages 643–644 or Transparencies
____ Daily Homework Quiz (10.7): TE page 648, or Transparencies
____ Standardized Test Practice: SE page 647; TE page 648; STP Workbook; Transparencies
____ Quiz (10.6–10.7): SE page 648

Notes ____

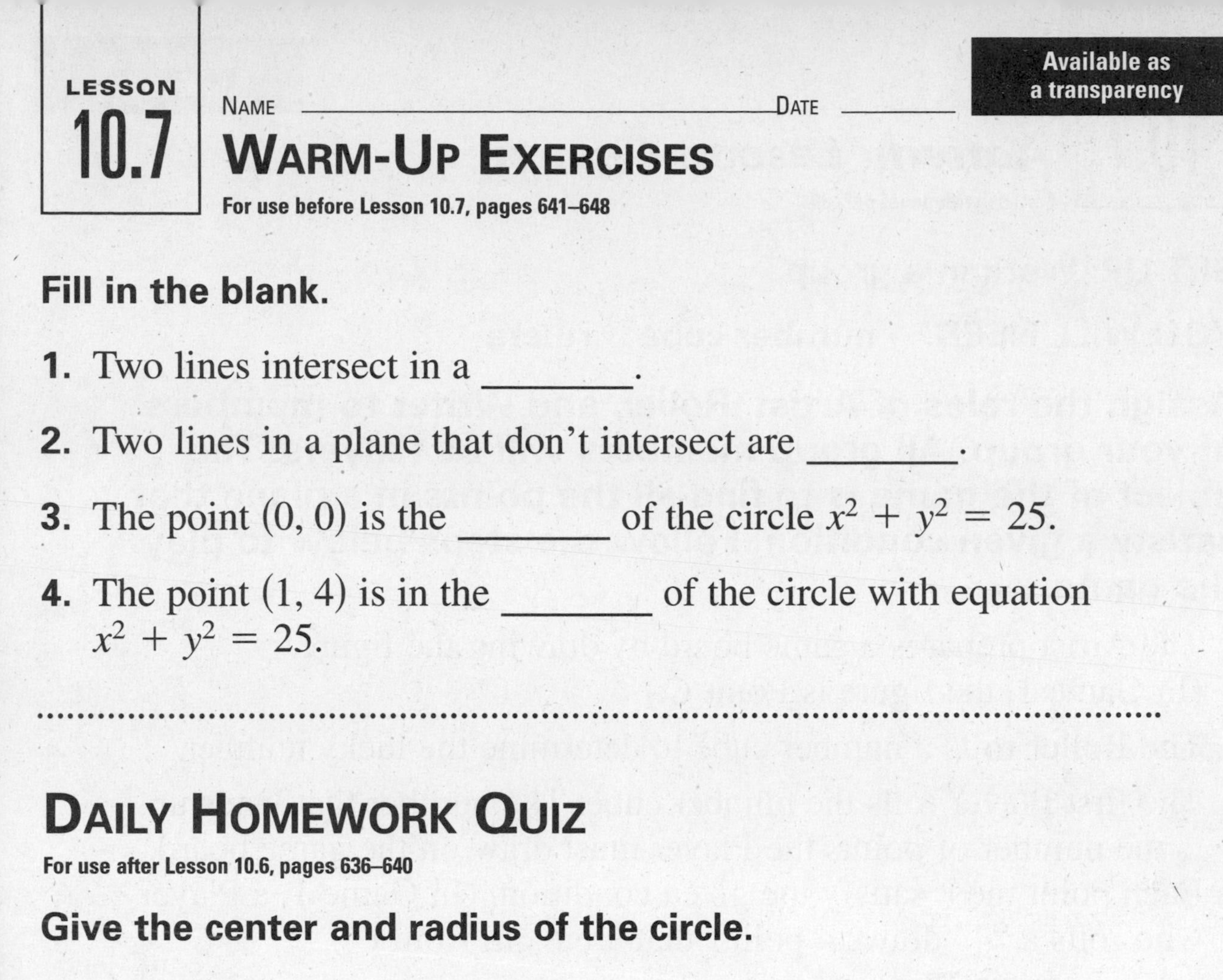

LESSON 10.7

NAME ______________________ DATE ________

Available as a transparency

WARM-UP EXERCISES

For use before Lesson 10.7, pages 641–648

Fill in the blank.

1. Two lines intersect in a ________.

2. Two lines in a plane that don't intersect are ________.

3. The point (0, 0) is the ________ of the circle $x^2 + y^2 = 25$.

4. The point (1, 4) is in the ________ of the circle with equation $x^2 + y^2 = 25$.

DAILY HOMEWORK QUIZ

For use after Lesson 10.6, pages 636–640

Give the center and radius of the circle.

1. $(x + 3)^2 + (y - 5)^2 = 36$

2. $(x - 4)^2 + (y + 1)^2 = 49$

Write the standard equation of the circle with the given information.

3. center (1, 0), radius 2

4. center (3, −2), a point on the circle is (3, 4).

Lesson 10.7

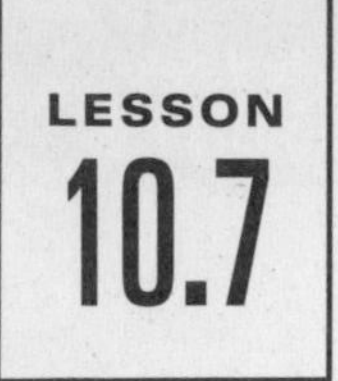

NAME ______________________ DATE __________

Available as a transparency

Activity Lesson Opener

For use with pages 642–648

SET UP: Work in a group.

YOU WILL NEED: • number cube • rulers

Assign the roles of Artist, Roller, and Writer to members of your group. All group members will be Players. The object of the game is to find all the points in a plane that satisfy a given condition. Follow the steps below to play the game.

- The Artist prepares a game board by drawing the figure. (In Game 1, the figure is Point C.)
- The Roller rolls a number cube to determine the lucky number.
- The first Player rolls the number cube. The number that lands up is the number of points the Player must draw on the game board. Each point must satisfy the given condition. (In Game 1, a Player who rolls a "4" draws 4 points that are 8 cm from C.)
- Players take turns rolling the cube and drawing points.
- The Player who rolls the lucky number must draw all points that satisfy the given condition. If the group agrees on this drawing, the Player wins. If not, play continues until someone wins.
- The Writer writes a sentence to describe the set of all points that satisfy the given condition, and the game is over.
- Switch roles for the next game.

Game 1 Figure: Point C
Points: 8 cm from C

Game 2 Figure: Point C
Points: Less than 8 cm from C

Game 3 Figure: Line k
Points: 2 inches from k

Game 4 Figure: Parallel lines m and n
Points: Equidistant from m and n

Game 5 Figure: Points A and B
Points: Equidistant from A and B

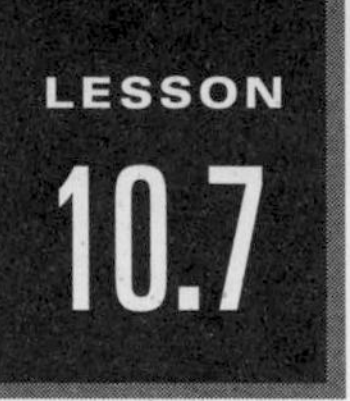

LESSON 10.7

NAME ______________________ DATE __________

Technology Activity Keystrokes

For use with page 641

TI-92

Construct

1. Draw line k using the line command (F2 4). Draw point P not on line k using the point command (F2 1). Draw $\overline{AB}$ in the corner of the screen using the segment command (F2 5).
2. Draw a line perpendicular to k.

   F4 1 (Place cursor on k.) ENTER (Move cursor off of k.) ENTER

   Label the point of intersection as point C using the intersection point command (F2 3). Construct a circle with center C and radius AB.

   F4 8 (Place cursor on C.) ENTER (Move cursor to $\overline{AB}$.) ENTER

   The circle intersects the line perpendicular to k in two points. Choose the point that is on the same side of k as P and label it D. Construct a line through D parallel to k. Label the line m.

   F4 2 (Place cursor on D.) ENTER (Move cursor to k.) ENTER m

   Hide $\overleftrightarrow{CD}$, $\odot C$, C, and D using the hide/show command (F7 1).
3. Construct a circle with center P and radius AB using the compass command (F4 8).
4. Draw the intersection points of circle P and line m and label the points Y and Z using the intersection point command (F2 3).

Investigate

2. Drag point B.

   F1 1 (Place cursor on B.) ENTER (Use the drag key [drag key] and the cursor pad to drag B.)
3. Use the *Trace* feature to trace Y and Z as you slowly drag B.

   F7 2 (Place cursor on Y.) ENTER (Move to unoccupied location in the plane.)

   ↑ + ENTER (Move cursor to Z.) ENTER (Move cursor to point B.)

   ENTER (Use the drag key [drag key] and the cursor pad to drag B.)

Extension

1. Turn on the axes and the grid.

   F8 9 (Set Coordinate Axes to RECTANGULAR and Grid to ON.) ENTER
2. Plot a point with a y-coordinate of -0.25 using the point command (F2 1).
3. Check the coordinates of the point.

   F6 5 (Place the cursor on the point.) ENTER (Adjust y-coordinate if necessary.)
4. Draw the line $y = -\frac{1}{4}$.

   F4 2 (Place cursor on plotted point.) ENTER (Move cursor to x-axis.) ENTER
5. Plot the point $\left(0, \frac{1}{4}\right)$ using the point command (F2 1).

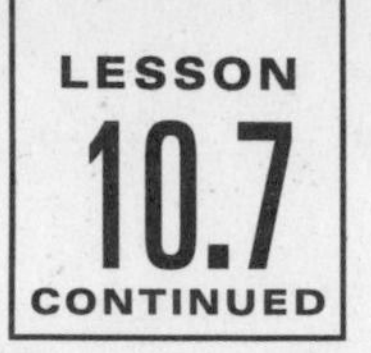

NAME ______________________________ DATE __________

Technology Activity Keystrokes

For use with page 641

SKETCHPAD

Construct

1. Draw line k using the line straightedge tool. Draw point P not on line k using the point tool. (Use the text tool to relabel the point.) Draw $\overline{AB}$ in the corner of the screen using the segment straightedge tool.
2. Construct point C on line k using the point tool. Draw a line perpendicular to k through C. Use the selection arrow tool to select k, hold down the shift key and select C, and choose **Perpendicular Line** from the **Construct** menu. Construct a circle with radius AB and center C. Use the selection arrow tool to select $\overline{AB}$, hold down the shift key and select C, and choose **Circle by Center and Radius** from the **Construct** menu. The circle intersects the line perpendicular to k in two points. Choose the point that is on the same side of k as P and label it D using the point and text tools. Construct a line through D parallel to k. Select D and k and choose **Parallel Line** from the **Construct** menu. Use the text tool to label the line m. Use the selection arrow tool to select $\overleftrightarrow{CD}$, $\odot C$, C, and D. Then choose **Hide Objects** from the **Display** menu.
3. Construct a circle with center P and radius AB. Select P and $\overline{AB}$, then choose **Circle by Center and Radius** from the **Construct** menu.
4. Draw the intersection points of circle P and line m using the point tool. Use the text tool to label the points Y and Z.

Lesson 10.7

Investigate

2. Use the translate selection arrow tool to drag point B.
3. Use the *Trace* feature to trace Y and Z as you slowly drag B. Use the selection arrow to select Y and Z. Choose **Trace Points** from the **Display** menu.

Extension

1. Turn on the axes and the grid. Choose **Snap to Grid** from the **Graph** menu.
2. Plot a point with a y-coordinate of -0.25. Choose **Plot Points** from the **Graph** menu. Enter an x-coordinate of your choice and a y-coordinate of 0.25.
3. Construct the line $y = -\frac{1}{4}$ using the line straightedge tool and the point from Extension Step 2.
4. Plot the point $\left(0, \frac{1}{4}\right)$ using the point tool.

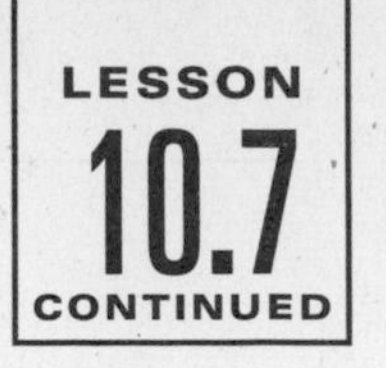

NAME ______________________ DATE __________

Technology Activity Keystrokes

For use with page 647

Keystrokes for Exercise 31

TI-92

1. Turn on the axes and the grid.

   F8 9 (Set Coordinate Axes to RECTANGULAR and Grid to ON.) ENTER

2. Draw a segment from $(-1, 1)$ to $(1, 1)$. Use the compass tool to draw a circle to represent the locus of points 2 units from P.

   F4 8 (Move cursor to segment.) ENTER (Move cursor to a blank area.)

   ENTER P

3. Use the line command to draw vertical lines through the following pairs of points: $(-1, 0)$ and $(-1, 1)$, $(1, 0)$ and $(1, 1)$, and $(3, 0)$ and $(3, 1)$. Label the second line k.

4. Drag P and notice the results.

   F1 1 (Move cursor to center of $\odot P$.) ENTER (Use the drag key and the cursor pad to drag the point.)

SKETCHPAD

1. Turn on the grid and axes by selecting **Snap to Grid** from the **Graph** menu.

2. Draw a segment from $(-1, 1)$ to $(1, 1)$ using the segment straightedge tool. Draw a point at $(-2, 1)$ using the point tool. Use the text tool to label the point at $(-2, 1)$ P. Use the selection arrow tool to select P and the segment. Choose **Circle by Center and Radius** from the **Construct** menu.

3. Use the line straightedge tool to draw vertical lines through the following pairs of points: $(-1, 0)$ and $(-1, 1)$, $(1, 0)$ and $(1, 1)$, and $(3, 0)$ and $(3, 1)$. Use the text tool to label the second line k.

4. Use the translate selection arrow tool to drag $\odot P$.

Lesson 10.7

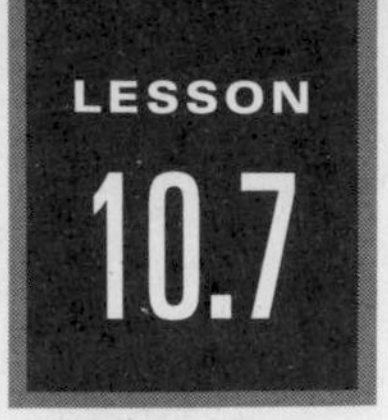

NAME ______________________ DATE ________

Practice A

For use with pages 642–648

Match the object with the locus of point *P*.

A. Circle **B.** Arc **C.** Line segment **D.** Parabola

1. **2.** **3.** **4.**

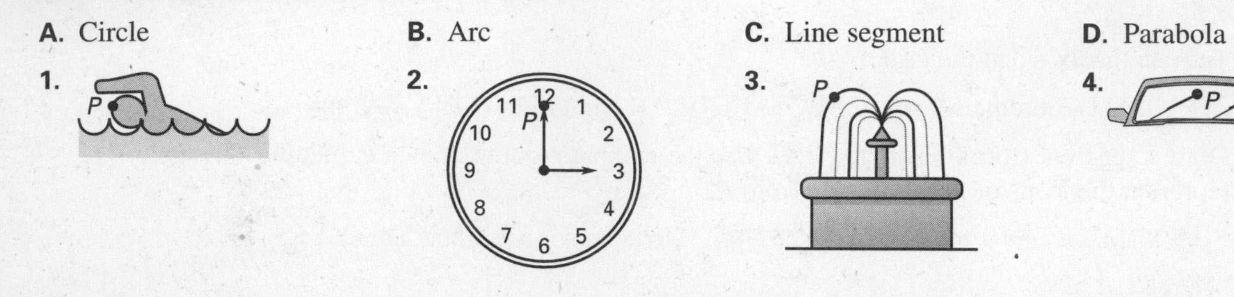

Match the sketch with the statement. Then describe the locus.

5. All points in a plane that are less than 1.2 centimeters from a given point

6. All points in a plane that are more than 1.2 centimeters from a given point

7. All points in a plane that are 1.2 centimeters or less from a given point

8. All points in a plane that are more than 0.8 centimeter and less than 1.2 centimeters from a given point

a.

b.

c.

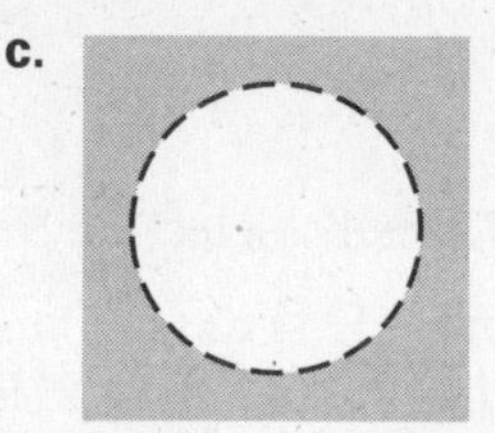

d.

Draw the figure. Then sketch and describe the locus points on the paper that satisfy the given conditions.

9. Right $\angle ABC$, the locus points on or in the interior of the angle and equidistant from the rays that form the angle

10. Points *A* and *B*, the locus of points that are equidistant from *A* and *B*

11. Line *l*, the locus of points that are 2 centimeters from *l*

12. Point *A*, the locus of points that are no more than 2 centimeters from *A*

Use the graph at the right to write the equation(s) for the locus of points in the coordinate plane that satisfy the given condition.

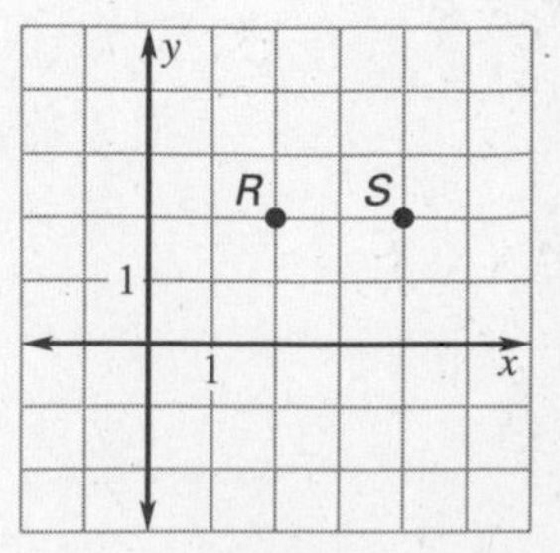

13. Equidistant from *R* and *S*

14. 2 units from *R*

15. Equidistant from the *x*- and *y*-axes

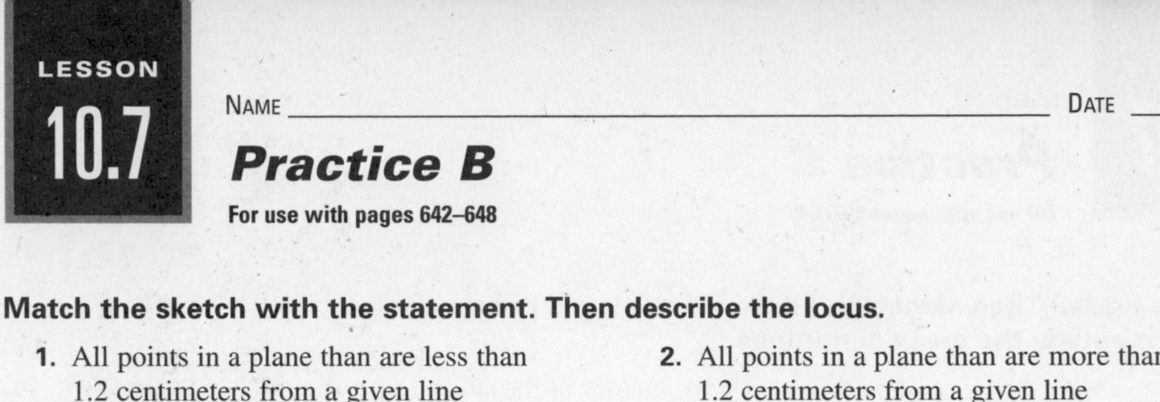

LESSON **10.7**

NAME ______________________________ DATE __________

Practice B

For use with pages 642–648

Match the sketch with the statement. Then describe the locus.

1. All points in a plane than are less than 1.2 centimeters from a given line
2. All points in a plane than are more than 1.2 centimeters from a given line
3. All points in a plane that are 1.2 centimeters or less from a given line
4. All points in a plane that are more than 0.8 centimeters and less than 1.2 centimeters from a given line

A. **B.**

C. **D.**

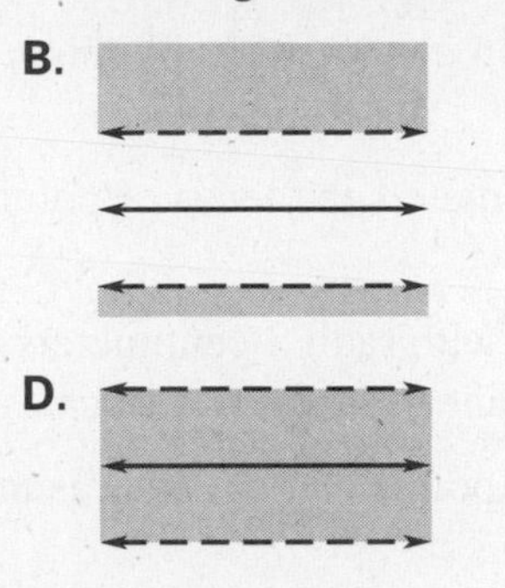

Draw the figure. Then sketch and describe the locus points on the paper that satisfy the given conditions.

5. Obtuse $\angle ABC$, the locus points on or in the interior of the angle and equidistant from the rays that form the angle
6. Square with side length 5, the locus points that are equidistant from the vertices of the square
7. Parallel lines m and l, the locus of the points that are equidistant from m and l
8. Circle of radius 2, the locus of points that are the midpoints of all radii of the circle

Use the graph at the right to write the equation(s) for the locus of points in the coordinate plane that satisfy the given condition.

9. Equidistant from R and S
10. 2 units from R
11. Equidistant from the x- and y-axes

12. ***Ceiling Fan*** An electrician is to install a ceiling fan in a rectangular room. It must be placed at a position which is equidistant from each of the four corners of the ceiling. Draw a diagram and describe the locus.
13. ***Flowers*** A gardner wishes to plant rows of flowers in a park. The flowers are to be equidistant from sidewalks that intersect as shown. Show the location of the flowers. Describe the locus of the flowers.

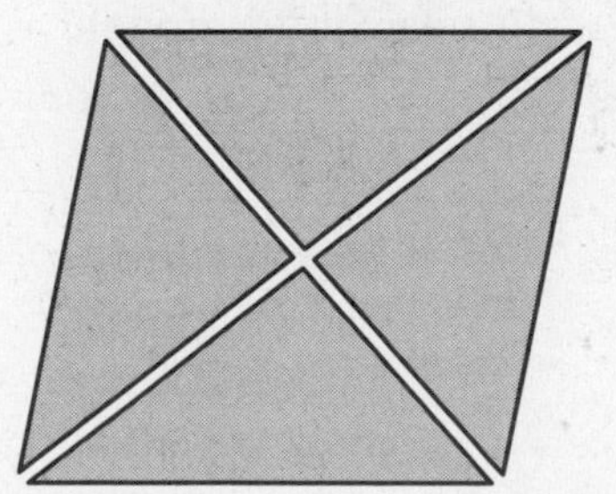

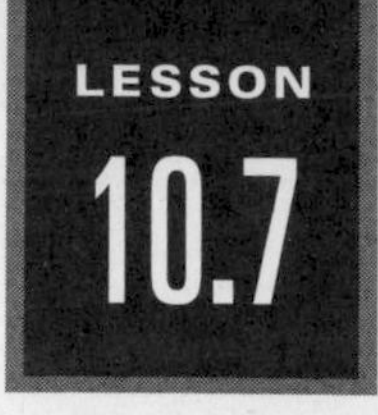

NAME ______________________________ DATE ____________

Practice C

For use with pages 642–648

Draw the figure. Then sketch and describe the locus points on the paper that satisfy the given conditions.

1. Acute $\angle ABC$, the locus of points on or in the interior of the angle and equidistant from the rays that form the angle
2. Segment $\overline{AB}$, all points that are equidistant from the endpoints of the segment
3. Segment $\overline{AB}$, of length 5 centimeters, all points that are 2 centimeters from the segment
4. Circle of radius 2 centimeters the locus of points that are 1 centimeter from the circle
5. Two concentric circles with radii 3 centimeters and 7 centimeters, all points that are equidistant from the two circles
6. Isosceles trapezoid, all points that are equidistant from the vertices of the isosceles trapezoid

Use the graph at the right to write the equation(s) for the locus of points in the coordinate plane that satisfy the given condition.

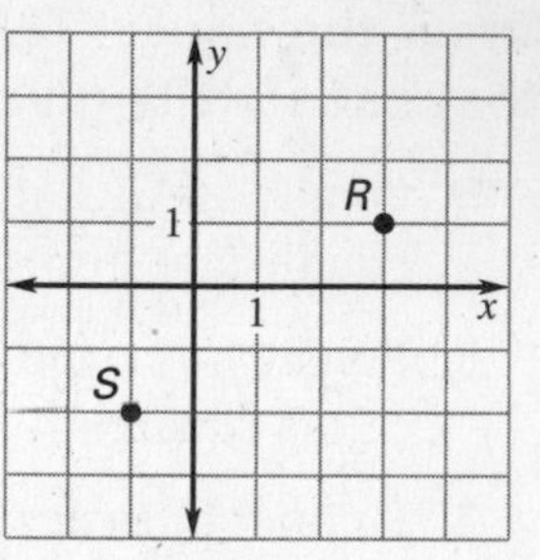

7. Equidistant from R and S
8. 2 units from R
9. Equidistant from the x- and y-axes

10. ***Coordinate Geometry*** Copy the triangle at the right. Construct the locus of points in a plane that are equidistant from the three vertices of the triangle.

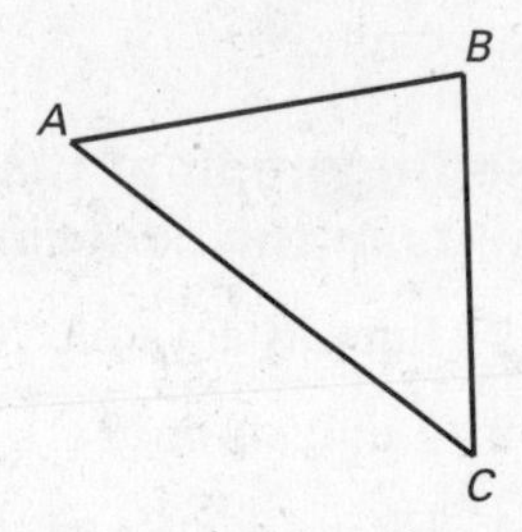

11. ***Coordinate Geometry*** Copy the triangle at the right. Construct the locus of points in a plane that are equidistant from the three sides of the triangle.

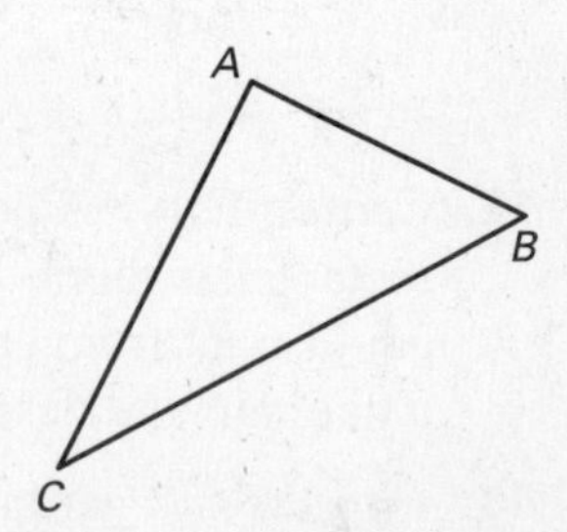

LESSON 10.7

NAME ______________________ DATE __________

Reteaching with Practice

For use with pages 642–648

GOAL **Draw the locus of points that satisfy a given condition and draw the locus of points that satisfy two or more conditions**

VOCABULARY

A **locus** in a plane is the set of all points in a plane that satisfy a given condition or set of given conditions.

Finding a Locus
To find the locus of points that satisfy a given condition, use the following steps.

1. Draw any figures that are given in the statement of the problem.
2. Locate several points that satisfy the given condition.
3. Continue drawing points until you can recognize the pattern.
4. Draw the locus and describe it in words.

EXAMPLE 1 *Finding a Locus*

Sketch and describe the locus of points that satisfy the given condition(s)

a. in the interior of $\angle P$ and equidistant from both sides of $\angle P$.

P

b. equidistant from j and k.

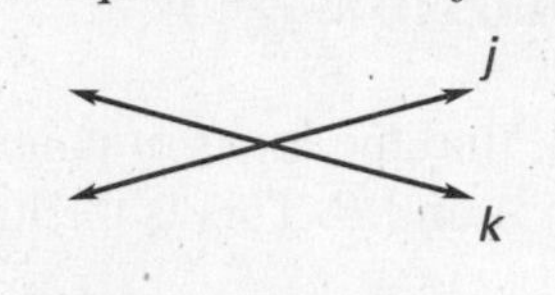

SOLUTION

a.

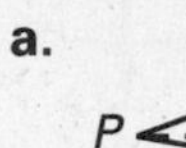

The points that make up the locus form the bisector of $\angle P$.

b.

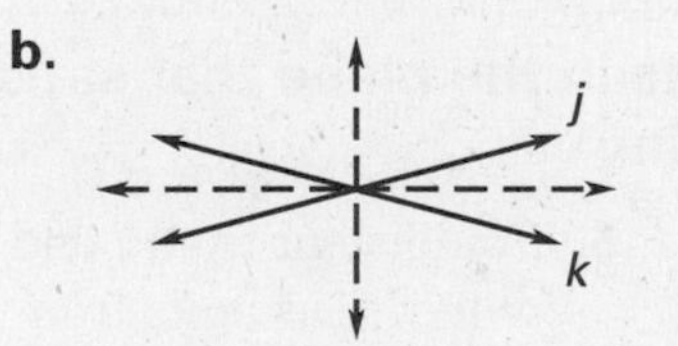

The points that make up the locus form the four angle bisectors of the angles formed by the intersection of the two lines.

Lesson 10.7

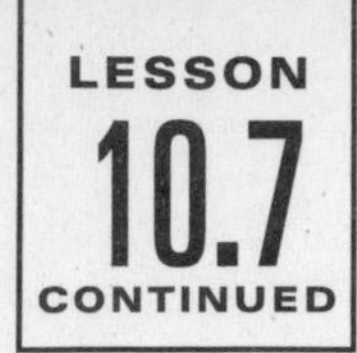

NAME ______________________________ DATE __________

Reteaching with Practice

For use with pages 642–648

Exercises for Example 1

Sketch and describe the locus of points that satisfy the given condition(s).

1. Equidistant from the lines $y = 3$ and $y = 7$
2. Within five units of the point $(-1, 2)$
3. Equidistant from $A(-1, 1)$ and $B(1, -1)$

EXAMPLE 2 Drawing a Locus Satisfying Two Conditions

Sketch and describe the locus of points in the plane that are equidistant from A and B and less than 3 units from the origin.

SOLUTION

First, find the locus of points that are equidistant from A and B. This is the line $x = 0$.

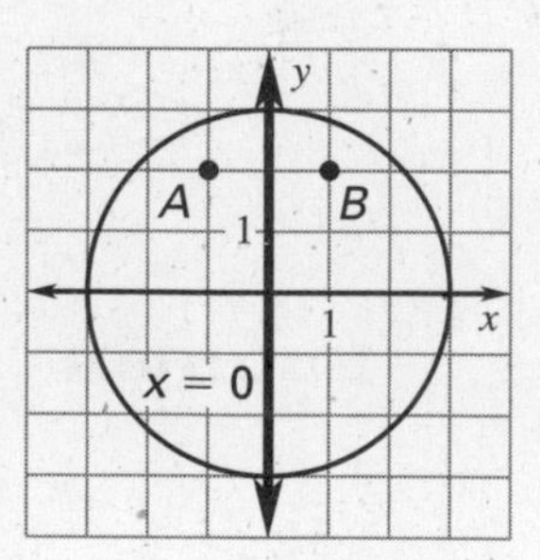

Next, find the locus of points that are less than 3 units from the origin. This is the circle centered at the origin with radius of 3.

Now, find the overlap of these two loci. This is the line segment from point $(0, 3)$ to $(0, -3)$.

Exercises for Example 2

Sketch and describe the locus of points in the plane that satisfy the given conditions. Explain your reasoning.

4. Equidistant from A and B and 2 units from the point $(1, 1)$.

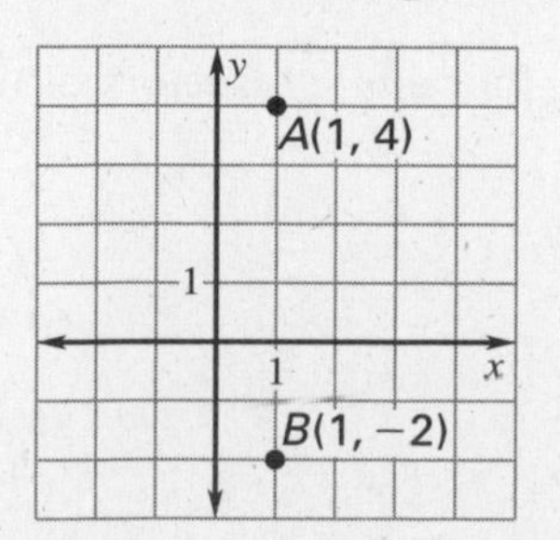

5. Equidistant from l and m and within four units from the origin.

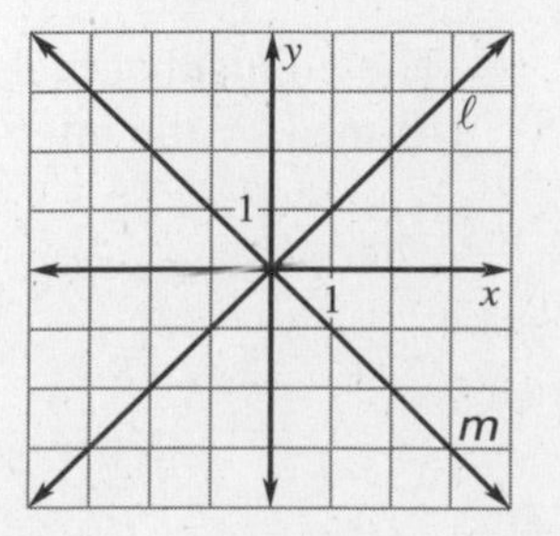

LESSON 10.7

NAME ______________________________ DATE __________

Quick Catch-Up for Absent Students

For use with pages 641–648

The items checked below were covered in class on (date missed) __________

Activity 10.7: Investigating Points Equidistant from a Point and a Line (p. 641)

____ **Goal:** Use geometry software to determine which points in a plane are equidistant from a point and a line in the plane.

Lesson 10.7: Locus

____ **Goal 1:** Draw the locus of points that satisfy a given condition. (p. 642)

Material Covered:

____ Example 1: Finding a Locus

Vocabulary:

locus, p. 642

____ **Goal 2:** Draw the locus of points that satisfy two or more conditions. (pp. 643–644)

Material Covered:

____ Example 2: Drawing a Locus Satisfying Two Conditions

____ Example 3: Drawing a Locus Satisfying Two Conditions

____ Example 4: Finding a Locus Satisfying Three Conditions

____ Other (specify) ______________________________

____

Homework and Additional Learning Support

____ Textbook (specify) pp. 576–580

____

____ *Reteaching with Practice* worksheet (specify exercises) __________

____ *Personal Student Tutor* for Lesson 10.7

LESSON 10.7

NAME ______________________________ DATE __________

Real-Life Application: When Will I Ever Use This?

For use with pages 642–648

Ropes Courses

A *ropes course* is a series of obstacles made of several different materials. A ropes course is designed to challenge an individual or group. People who go through a ropes course work their way up from low course elements (up to three feet off the ground) to high course elements (20–85 feet off the ground). Using teamwork and rock climbing equipment, the course provides a practically risk free controlled situation. However, to the participant, it does not feel like it is risk free. This is what makes a ropes course so beneficial. A ropes course builds confidence and team problem solving. Young and old alike can participate in a ropes course. Companies often provide such an experience to employees to encourage teamwork, which in turn builds a better business.

In Exercises 1–4, use the following information and diagram.

You and a group of friends are participating in a ropes course. One of the obstacles involves doing a Tarzan Swing into a lake. A rope swing is at the edge of the lake. It is attached to a platform arm that overhangs the lake as shown in the diagram below. The point at which the rope is attached is 30 feet above the water and 20 feet from the bank. The platform is on the 20-foot bank at the edge of the lake and is 20 feet high. The rope is 35 feet long, allowing the rope to hang down in the water so that a swimmer can reach it and tow it back to shore for the next group member. The rope is knotted every five feet so that swingers have something to hold as they swing out over the lake.

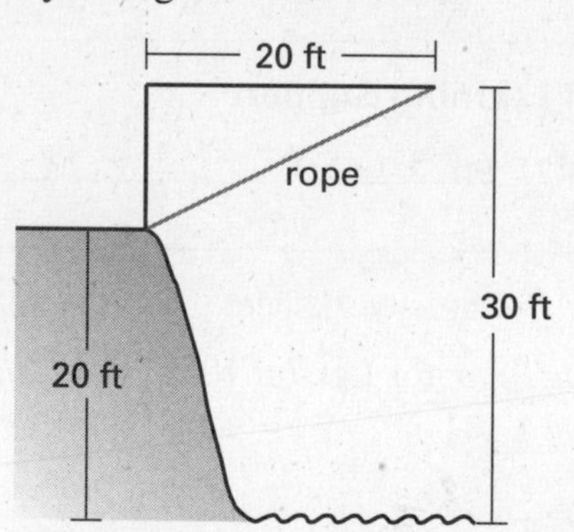

1. How long is the part of the rope that is stretched from the attachment point to the edge of the bank?
2. Where should you grab the rope so that you can clear the bank and land safely in the water?
3. How high above the water will you be when the rope is straight down?
4. Sketch the locus of points that will show your swing pattern if you swing out 10 feet beyond vertical before you let go.

Lesson 10.7

NAME ______________________ DATE __________

Math and History Application

For use with page 648

MATH The ancient inhabitants of Babylonia (3000 B.C.–A.D. 260) used a *sexagesimal* or base 60 number system; that is, the Babylonians represented each number as the sum of powers of 60. For example,

$$7296 = 2 \cdot 60^2 + 1 \cdot 60 + 36 \qquad (1)$$

The same was also done with fractions; each fraction was written as the sum of powers of $\frac{1}{60}$. For example,

$$\frac{3}{8} = 22\left(\frac{1}{60}\right) + 30\left(\frac{1}{60}\right)^2 \qquad (2)$$

HISTORY Through trade and conquest, Alexander the Great (356–323 B.C.) extended his empire to the limits of the known world, including Babylonia. Soon afterwards, Greek astronomers adopted the Babylonian sexagesimal system to record their own data. Fractions of the form $\frac{1}{60}$ were called "the first small parts," fractions of the form $\left(\frac{1}{60}\right)^2$ were called "the second small parts," and so on.

Centuries later, when Greek astronomical data was translated into Latin, "the first small parts" was translated as *pars minuta prima*, and "the second small parts" was translated as *pars minuta secunda*. When these phrases, in turn, were translated into English, they became *minutes* and *seconds*, respectively.

1. Verify Equations (1) and (2) above.

2. Write each number in base 60 notation.

a. 588 **b.** 4406 **c.** $\frac{3}{4}$

3. Write each sum in base 10 notation.

a. $20 \cdot 60^2 + 21 \cdot 60 + 44$

b. $3 + 7\left(\frac{1}{60}\right) + 30\left(\frac{1}{60}\right)^2$ (Babylonian approximation of π)

4. Explain why you think the Babylonians used a base 60 number system.

NAME ______________________ DATE ____________

Challenge: Skills and Applications

For use with pages 642–648

In Exercises 1–4, sketch and describe the locus points on the paper that satisfy the given condition(s).

1. The locus of points that are equidistant from the three sides of the triangle

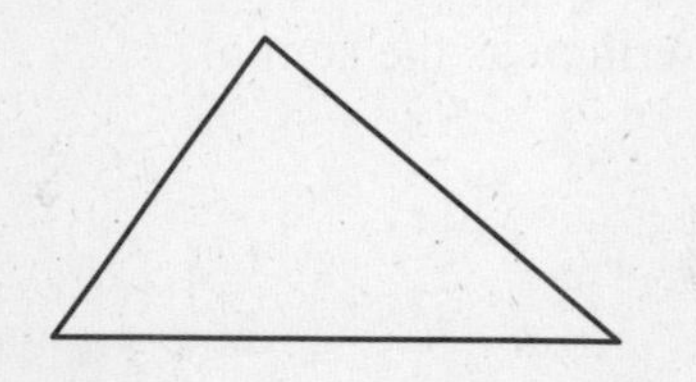

2. The locus of points that are equidistant from the four vertices of the nonrectangular parallelogram

3. Given three vertices of a parallelogram, the locus of points that could be the fourth vertex

R

P

Q

4. The locus of points that are a distance of 5 mm from the circle (That is, the distance to the *nearest* point on the circle is 5 mm.)

5. Given two parallel planes in space, describe the locus of points that are equidistant from the two planes.

6. Given a line in space, describe the locus of points whose distance from the line is 3 inches.

7. Given a plane in space, describe the locus of points whose distance from the plane is 6 centimeters.

8. Given two points in space, describe the locus of points that are equidistant from the two points.

9. Let C be a point in space, and let r be a positive number. A *sphere* with center C and radius r can be described as the locus of points whose distance from C is r. If S is a sphere centered at the origin with radius 3, describe the locus of points whose distance from S is 4.

NAME ______________________ DATE __________

Chapter Review Games and Activities

For use after Chapter 10

Fill in each blank. Each answer is a positive integer. Then put your first six answers in the spaces provided in diagram 1 in such a way that the sum of all the numbers along each circle is 22. Do the same for your last six answers and diagram two.

1. If the radius of a circle is 5, its diameter is__________.

2. A secant is a line that intersects a circle in __________ points.

3. A tangent is a line in the plane of a circle that intersects the circle in __________ point(s).

4. Suppose that R and T are points on a circle and that S is a point exterior to the circle. Suppose also that $\overline{SR}$ and $\overline{ST}$ are tangent to the circle. The length of $\overline{SR}$ is 8. The length of $\overline{ST}$ is __________.

5. Suppose that points A, B, and C are on the same circle. Suppose also that $\widehat{AB} \cong \widehat{BC}$ and that the length of $\overline{AB} = 3$. The length of $\overline{BC}$ is __________.

6. Suppose that points A, B, and C are on the same circle and that $m\widehat{AB}$ is 18°. Then $m\angle ACB$ is __________.

7. A right triangle is inscribed in a circle of radius 3. The length of the hypotenuse of the right triangle is __________.

8. Suppose that the points A, B, C, and D, in that order, are on the same circle and that $m\angle ABC = 174°$. Then $m\angle CDA$ is __________.

9. Suppose that A and B are points on a circle and C is an exterior point such that $\overleftrightarrow{AC}$ is tangent to the circle. If $m\widehat{AB}$ is 10°, then $m\angle BAC$ is __________.

10. Suppose that A, B, C, and D, in that order, are points on a circle with center P. Suppose also that $m\widehat{AB}$ is 6° and that $m\widehat{CD}$ is 4°. Then $m\angle APB$ is __________.

11. The equation of a circle of radius 2 centered at (0, 0) is $x^2 + y^2 =$ __________.

12. Suppose that the points A, B, C, and D, in that order, are on the same circle. Suppose also that $\overline{AC}$ intersects $\overline{BD}$ at the point E. If the length of $\overline{BE} = 2$, the length of $\overline{DE} = 3$, and the length of $\overline{AE} = 1$, then the length of $\overline{AC} =$ __________.

Diagram 1

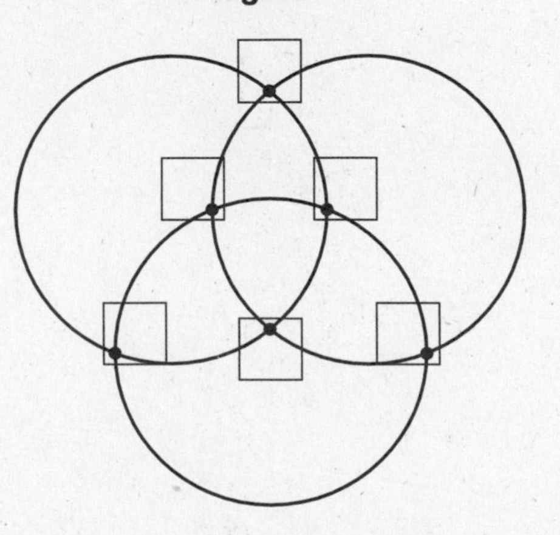

Diagram 2

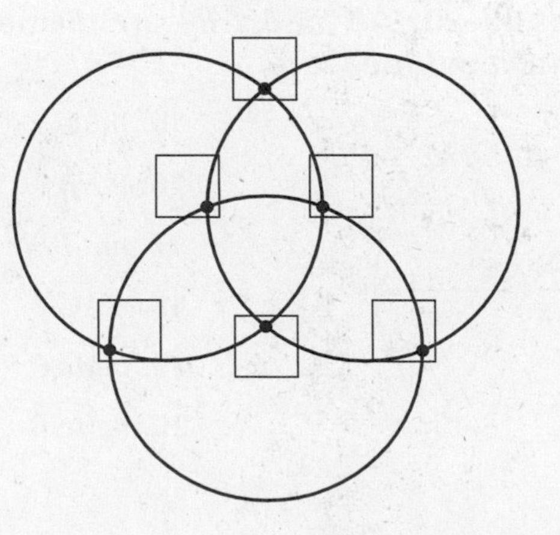

Review and Assess

CHAPTER 10

NAME ______________________ DATE __________

Chapter Test A

For use after Chapter 10

The diameter of a circle is given. Find the radius.

1. $d = 8$ ft **2.** $d = 9$ cm **3.** $d = 2.1$ m

The radius of ⊙*B* is given. Find the diameter of ⊙*B*.

4. $r = 21$ cm **5.** $r = 33$ ft **6.** $r = 2.9$ m

Using the diagram below, match the notation with the term that best describes it.

7. Chord	**A.** $\overline{CB}$
8. Point of tangency	**B.** $\overleftrightarrow{BG}$
9. Common Internal Tangent	**C.** $\overleftrightarrow{DE}$
10. Common External Tangent	**D.** A
11. Center	**E.** $\overline{EB}$
12. Radius	**F.** $\overleftrightarrow{AH}$
13. Diameter	**G.** $\overline{DE}$
14. Secant	**H.** C

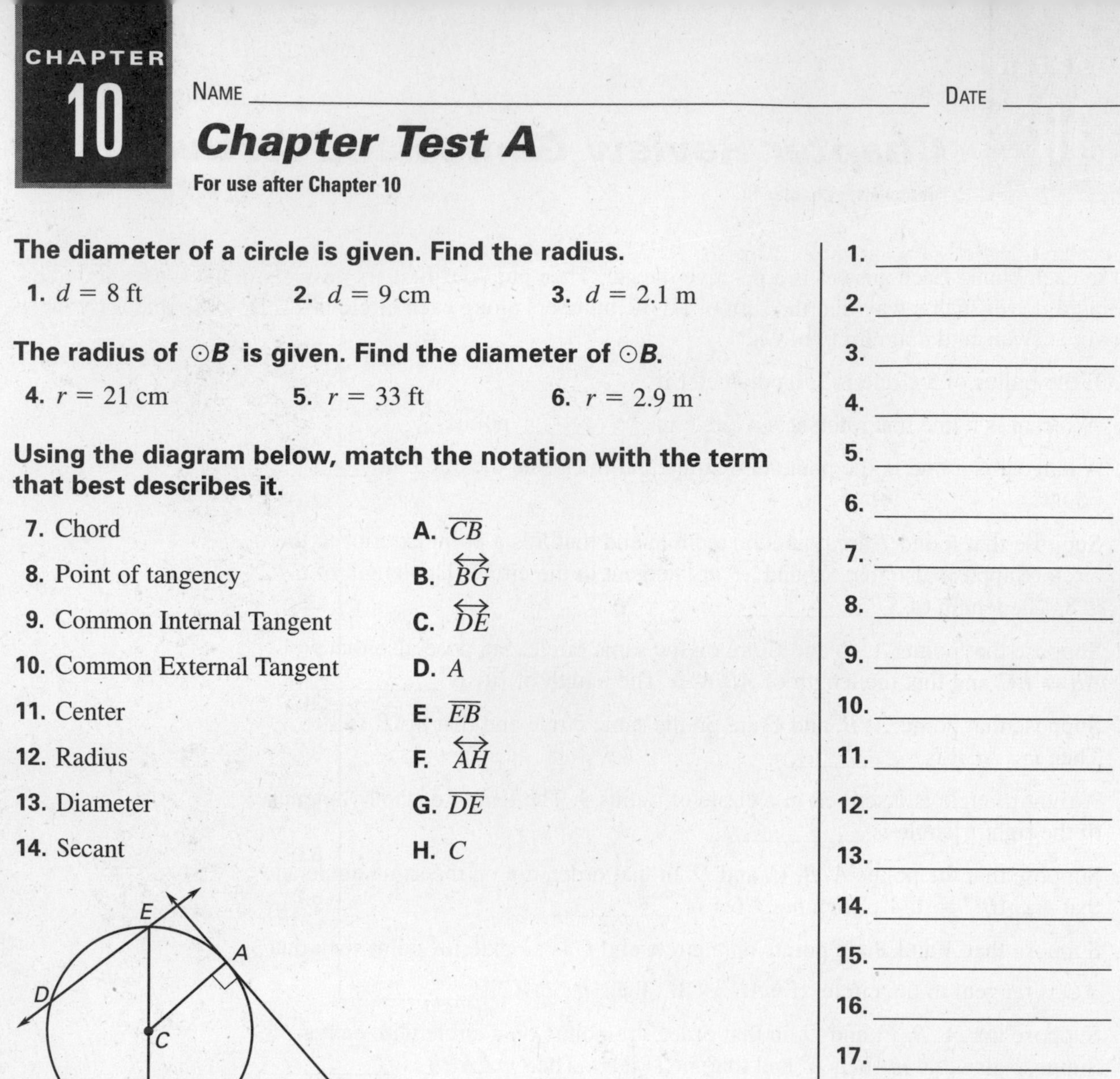

1. ______
2. ______
3. ______
4. ______
5. ______
6. ______
7. ______
8. ______
9. ______
10. ______
11. ______
12. ______
13. ______
14. ______
15. ______
16. ______
17. ______
18. ______
19. ______
20. ______

In Exercises 15–20, $\overline{AB}$ and $\overline{MN}$ are diameters of ⊙*E*. Find the indicated measure.

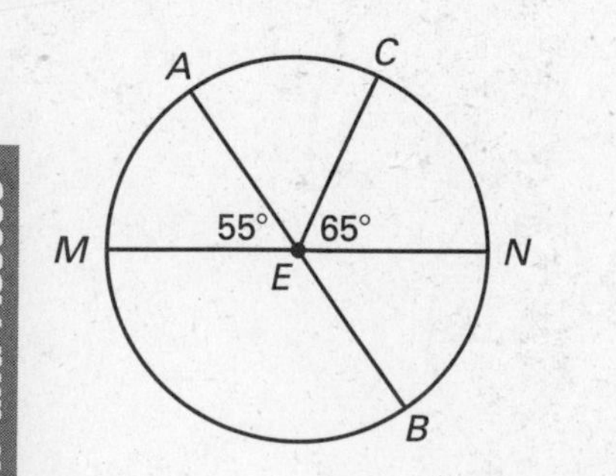

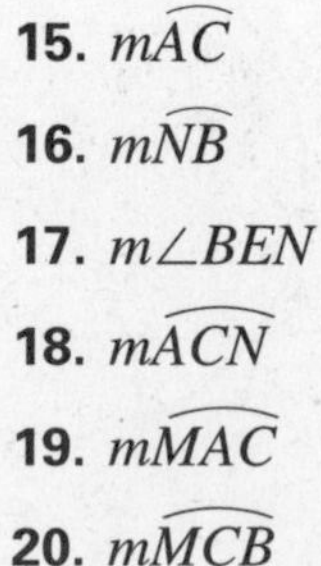

15. $m\overset{\frown}{AC}$

16. $m\overset{\frown}{NB}$

17. $m\angle BEN$

18. $m\overset{\frown}{ACN}$

19. $m\overset{\frown}{MAC}$

20. $m\overset{\frown}{MCB}$

Review and Assess

CHAPTER 10 CONTINUED

NAME ______________________ DATE __________

Chapter Test A

For use after Chapter 10

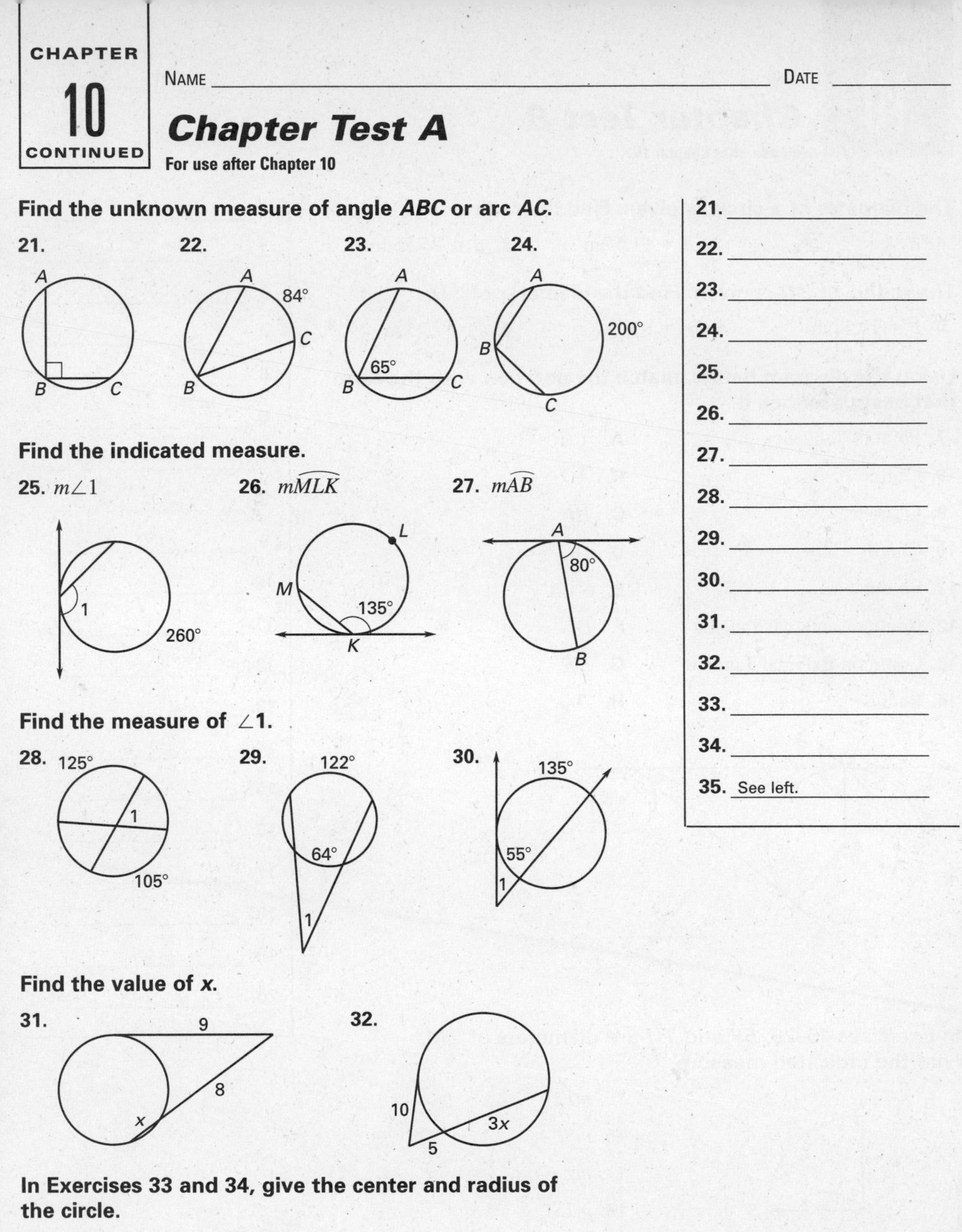

Find the unknown measure of angle *ABC* or arc *AC*.

21.

22.

23.

24.

Find the indicated measure.

25. $m\angle 1$

26. $m\widehat{MLK}$

27. $m\widehat{AB}$

Find the measure of $\angle$1.

28.

29.

30.

Find the value of *x*.

31.

32.

21. ______
22. ______
23. ______
24. ______
25. ______
26. ______
27. ______
28. ______
29. ______
30. ______
31. ______
32. ______
33. ______
34. ______
35. See left.

In Exercises 33 and 34, give the center and radius of the circle.

33. $(x + 3)^2 + (y - 4)^2 = 36$

34. $(x + 1)^2 + (y - 6)^2 = 2.25$

35. Describe the locus of points in a plane 3 inches from a given point C.

____

____

Review and Assess

CHAPTER 10

NAME ______________________ DATE __________

Chapter Test B

For use after Chapter 10

The diameter of a circle is given. Find the radius.

1. $d = 15$ ft
2. $d = 11.5$ cm
3. $d = 25.25$ in.

The radius of $\odot D$ is given. Find the diameter of $\odot D$.

4. $r = 10.5$ cm
5. $r = 100$ in.
6. $r = 7.75$ m

Using the diagram below, match the notation with the term that best describes it.

7. Point of Tangency
8. Center
9. Diameter
10. Chord
11. Secant
12. Common Internal Tangent
13. Common External Tangent
14. Radius

A. $\overleftrightarrow{CH}$
B. $\overline{AD}$
C. $\overline{HF}$
D. $\overleftrightarrow{DG}$
E. C
F. $\overline{DG}$
G. $\overleftrightarrow{DE}$
H. B

In Exercises 15–20, $\overline{EF}$ and $\overline{TU}$ are diameters of $\odot M$. Find the indicated measure.

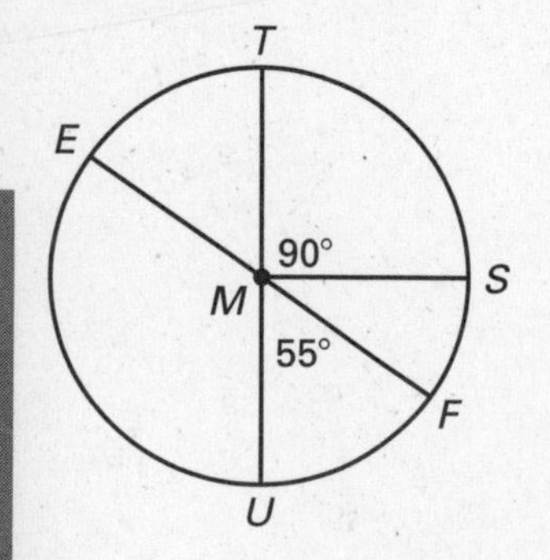

15. $m\overset{\frown}{ET}$
16. $m\overset{\frown}{SF}$
17. $m\angle EMS$
18. $m\overset{\frown}{TSF}$
19. $m\angle SMU$
20. $m\angle EMU$

1. ______
2. ______
3. ______
4. ______
5. ______
6. ______
7. ______
8. ______
9. ______
10. ______
11. ______
12. ______
13. ______
14. ______
15. ______
16. ______
17. ______
18. ______
19. ______
20. ______

Review and Assess

CHAPTER 10 CONTINUED

NAME ______________________________ DATE __________

Chapter Test B

For use after Chapter 10

Find the unknown measure of angle *LMN* or arc *LN*.

21.

22.

23.

24.

Find the indicated measure.

25. $m\widehat{AB}$

26. $m\angle 1$

27. $m\angle 5$

Find the measure of $\angle 1$.

28.

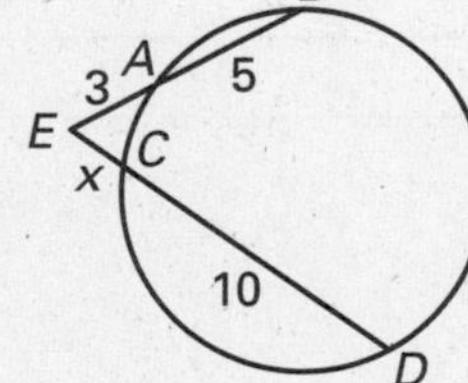

29.

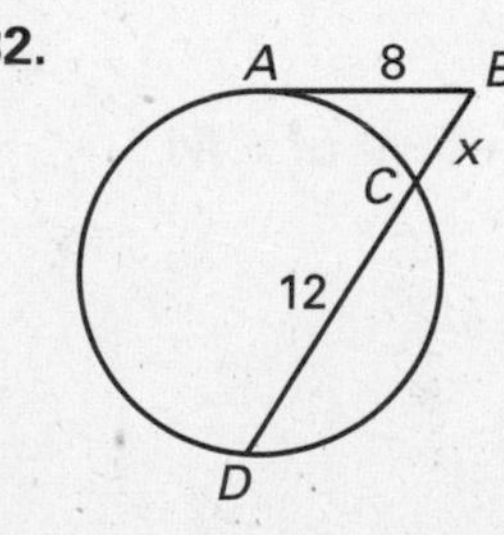

30.

Find the value of *x*.

31.

32.

In Exercises 33 and 34, give the center and radius of the circle.

33. $(x + 2)^2 + (y + 2)^2 = 25$

34. $\left(x + \frac{1}{2}\right)^2 + (y - 6)^2 = 35$

35. Describe the locus of points in a plane 5 centimeters from a given point D.

____

____

21. ________
22. ________
23. ________
24. ________
25. ________
26. ________
27. ________
28. ________
29. ________
30. ________
31. ________
32. ________
33. ________
34. ________
35. See left.

Review and Assess

CHAPTER 10

NAME ______________________________ DATE ____________

Chapter Test C

For use after Chapter 10

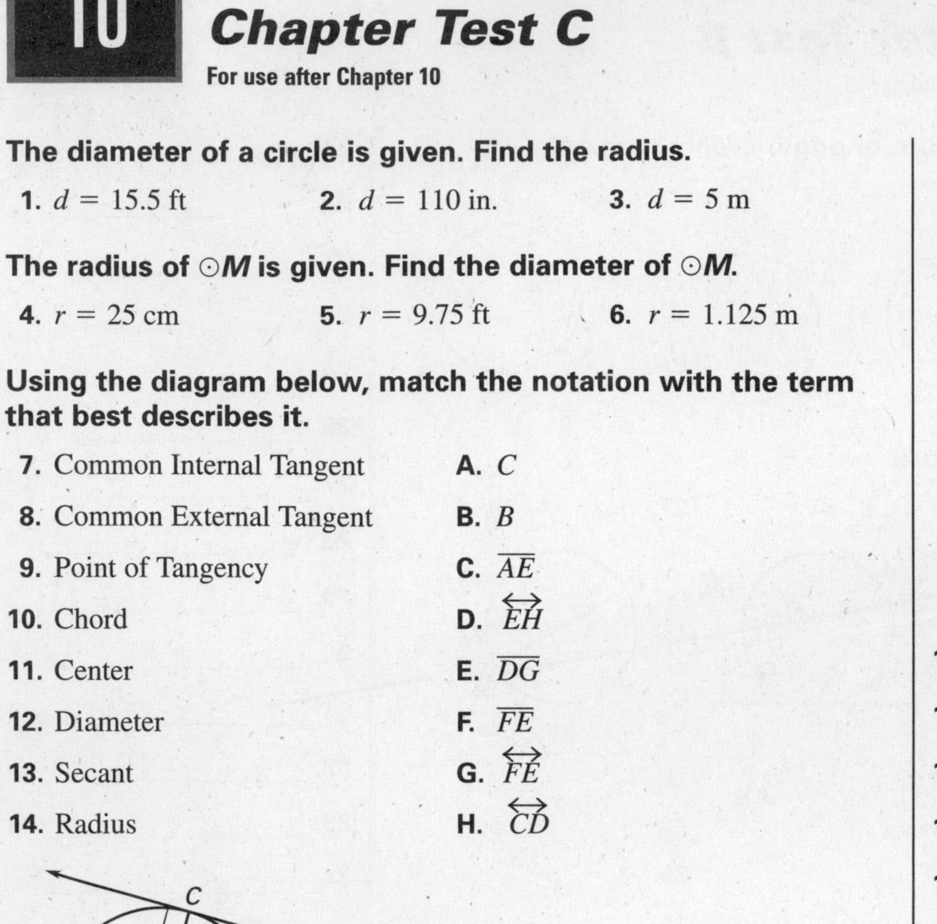

The diameter of a circle is given. Find the radius.

1. $d = 15.5$ ft
2. $d = 110$ in.
3. $d = 5$ m

The radius of $\odot M$ is given. Find the diameter of $\odot M$.

4. $r = 25$ cm
5. $r = 9.75$ ft
6. $r = 1.125$ m

Using the diagram below, match the notation with the term that best describes it.

7. Common Internal Tangent	A. C
8. Common External Tangent	B. B
9. Point of Tangency	C. $\overline{AE}$
10. Chord	D. $\overleftrightarrow{EH}$
11. Center	E. $\overline{DG}$
12. Diameter	F. $\overline{FE}$
13. Secant	G. $\overleftrightarrow{FE}$
14. Radius	H. $\overleftrightarrow{CD}$

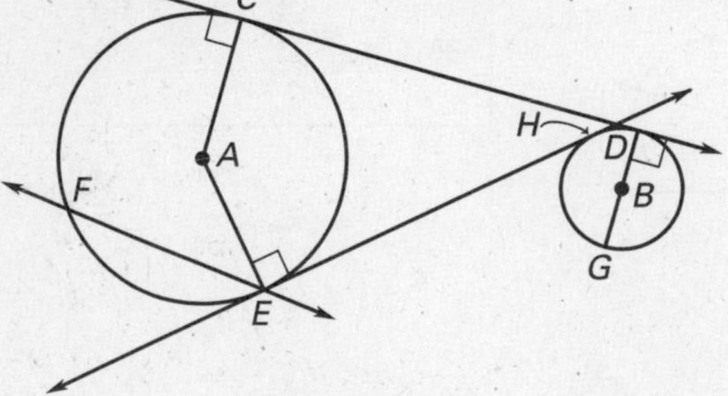

In Exercises 15–20, $\overline{KL}$ and $\overline{XY}$ are diameters of $\odot M$. Find the indicated measure.

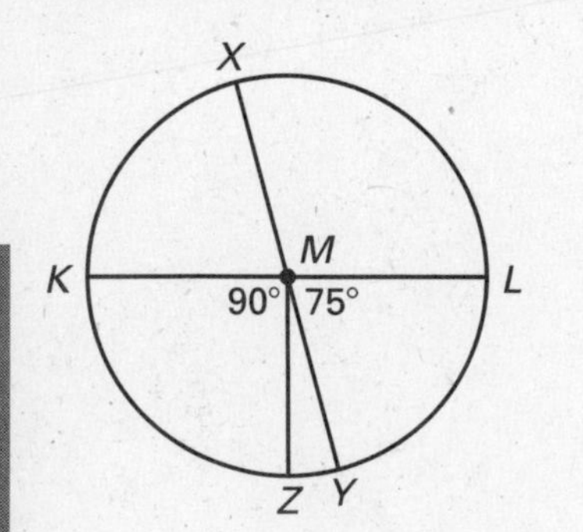

15. $m\angle KMX$
16. $m\overparen{YZ}$
17. $m\overparen{ZYL}$
18. $m\angle XML$
19. $m\overparen{XLZ}$
20. $m\overparen{KXY}$

1. ________
2. ________
3. ________
4. ________
5. ________
6. ________
7. ________
8. ________
9. ________
10. ________
11. ________
12. ________
13. ________
14. ________
15. ________
16. ________
17. ________
18. ________
19. ________
20. ________

Review and Assess

NAME ______________________________ DATE __________

Chapter Test C

For use after Chapter 10

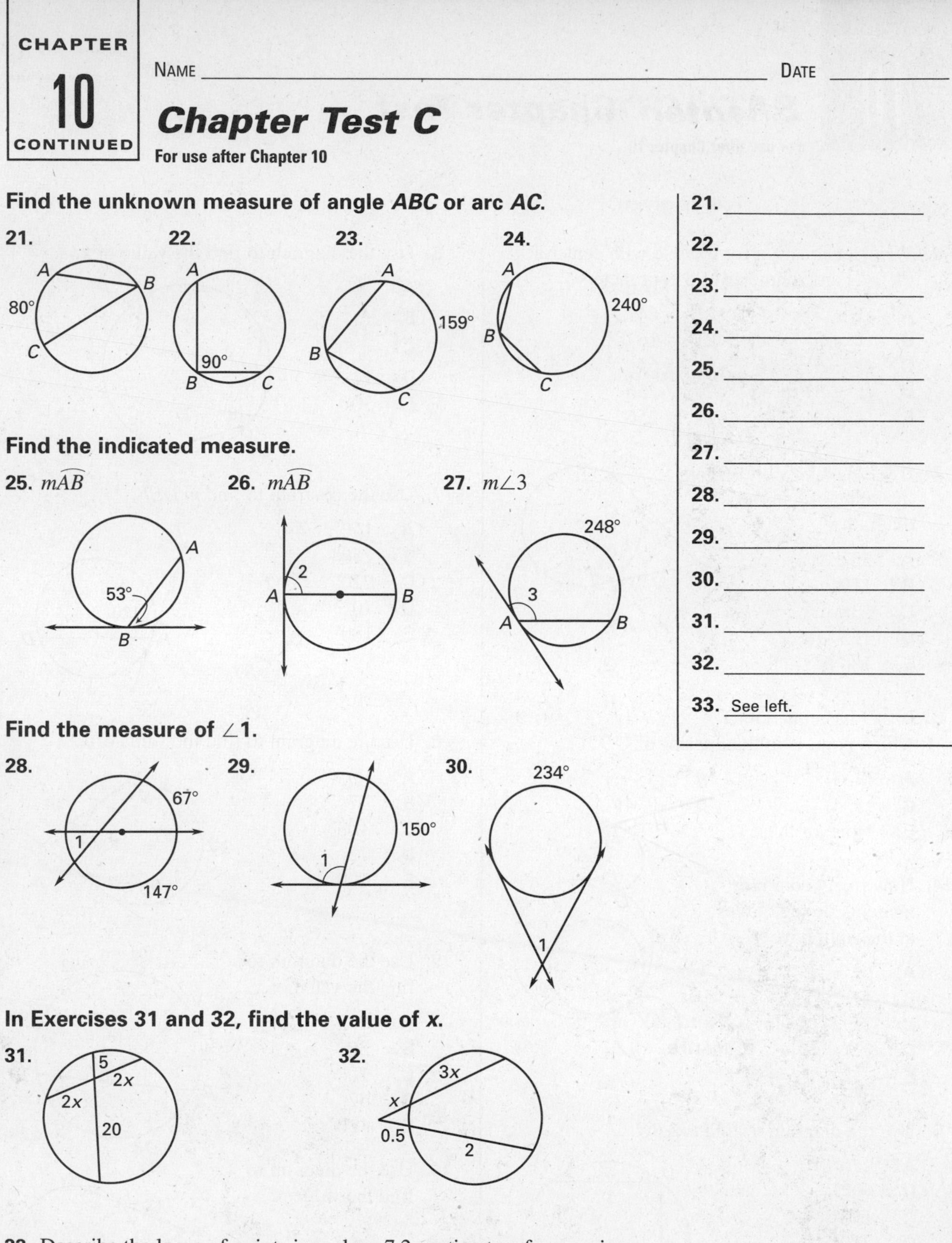

Find the unknown measure of angle *ABC* or arc *AC*.

21.

22.

23.

24.

Find the indicated measure.

25. $m\widehat{AB}$

26. $m\widehat{AB}$

27. $m\angle 3$

Find the measure of $\angle 1$.

28.

29.

30.

In Exercises 31 and 32, find the value of *x*.

31.

32.

21. ____________
22. ____________
23. ____________
24. ____________
25. ____________
26. ____________
27. ____________
28. ____________
29. ____________
30. ____________
31. ____________
32. ____________
33. See left.

33. Describe the locus of points in a plane 7.2 centimeters from a given point *A*.

____

____

Review and Assess

NAME ______________________ DATE __________

SAT/ACT Chapter Test

For use after Chapter 10

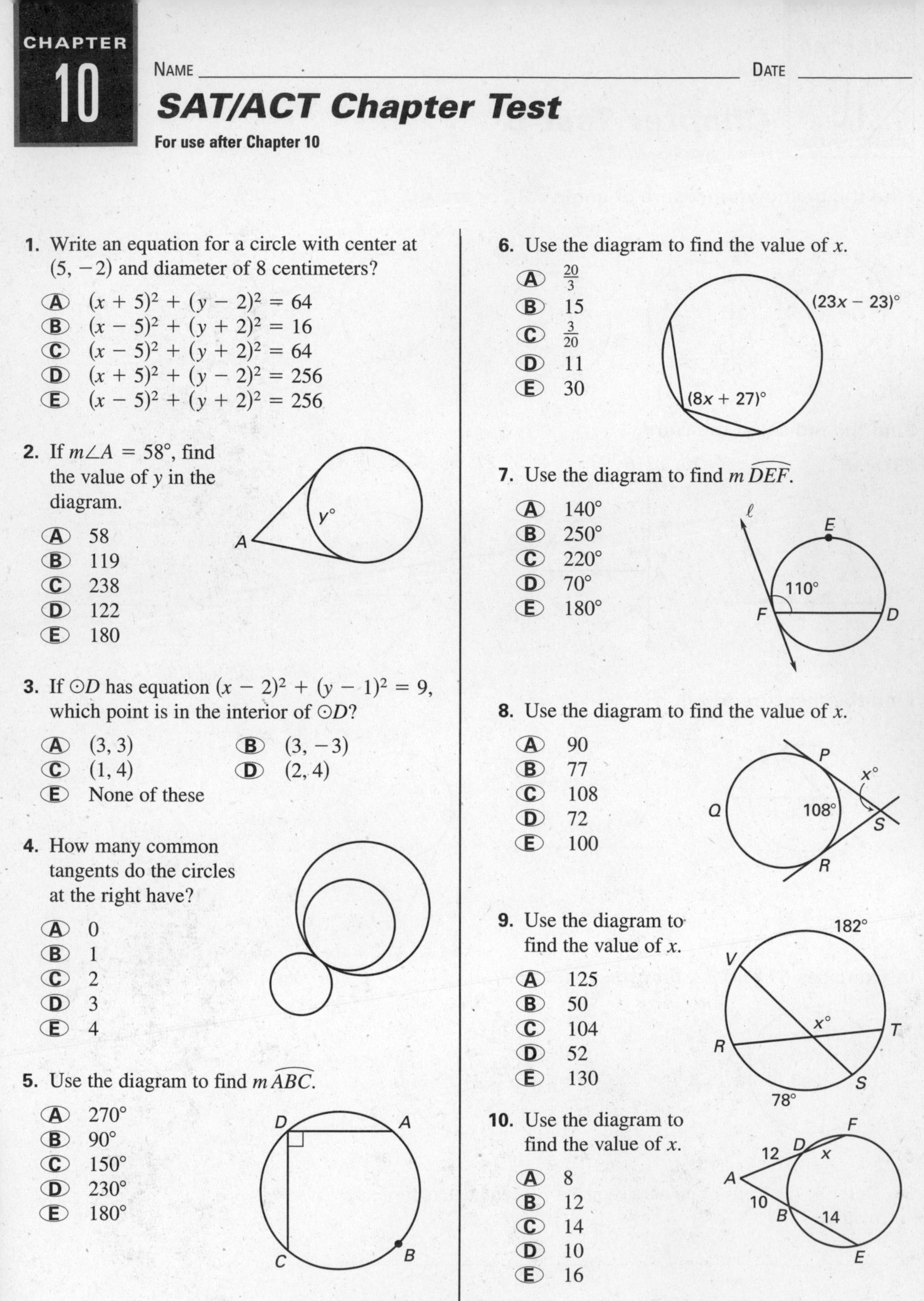

1. Write an equation for a circle with center at $(5, -2)$ and diameter of 8 centimeters?

Ⓐ $(x + 5)^2 + (y - 2)^2 = 64$
Ⓑ $(x - 5)^2 + (y + 2)^2 = 16$
Ⓒ $(x - 5)^2 + (y + 2)^2 = 64$
Ⓓ $(x + 5)^2 + (y - 2)^2 = 256$
Ⓔ $(x - 5)^2 + (y + 2)^2 = 256$

2. If $m\angle A = 58°$, find the value of y in the diagram.

Ⓐ 58
Ⓑ 119
Ⓒ 238
Ⓓ 122
Ⓔ 180

3. If $\odot D$ has equation $(x - 2)^2 + (y - 1)^2 = 9$, which point is in the interior of $\odot D$?

Ⓐ $(3, 3)$ Ⓑ $(3, -3)$
Ⓒ $(1, 4)$ Ⓓ $(2, 4)$
Ⓔ None of these

4. How many common tangents do the circles at the right have?

Ⓐ 0
Ⓑ 1
Ⓒ 2
Ⓓ 3
Ⓔ 4

5. Use the diagram to find $m\widehat{ABC}$.

Ⓐ 270°
Ⓑ 90°
Ⓒ 150°
Ⓓ 230°
Ⓔ 180°

6. Use the diagram to find the value of x.

Ⓐ $\frac{20}{3}$
Ⓑ 15
Ⓒ $\frac{3}{20}$
Ⓓ 11
Ⓔ 30

7. Use the diagram to find $m\widehat{DEF}$.

Ⓐ 140°
Ⓑ 250°
Ⓒ 220°
Ⓓ 70°
Ⓔ 180°

8. Use the diagram to find the value of x.

Ⓐ 90
Ⓑ 77
Ⓒ 108
Ⓓ 72
Ⓔ 100

9. Use the diagram to find the value of x.

Ⓐ 125
Ⓑ 50
Ⓒ 104
Ⓓ 52
Ⓔ 130

10. Use the diagram to find the value of x.

Ⓐ 8
Ⓑ 12
Ⓒ 14
Ⓓ 10
Ⓔ 16

NAME ______________________ DATE __________

Alternative Assessment and Math Journal

For use after Chapter 10

JOURNAL **1.** Draw a circle O with a diameter of 3 inches. Label the horizontal diameter $\overline{AC}$. Draw radius $\overline{OF}$ to form a minor arc $\overset{\frown}{CF}$. Draw a tangent to circle O at point C. Find the measure of the angle formed by the tangent and $\overline{AC}$. Draw chord $\overline{MN}$. Draw secant $\overleftrightarrow{AF}$. Using circle O, add another circle so the two circles are concentric.

MULTI-STEP PROBLEM **2.** You are swimming in a circular pool shown at the right. The ladder is at C. You are standing in the middle of the pool at M. Your friend is standing next to the side of the pool at B. There is a basketball hoop at A. The measure of $\angle CMB$ is 114°.

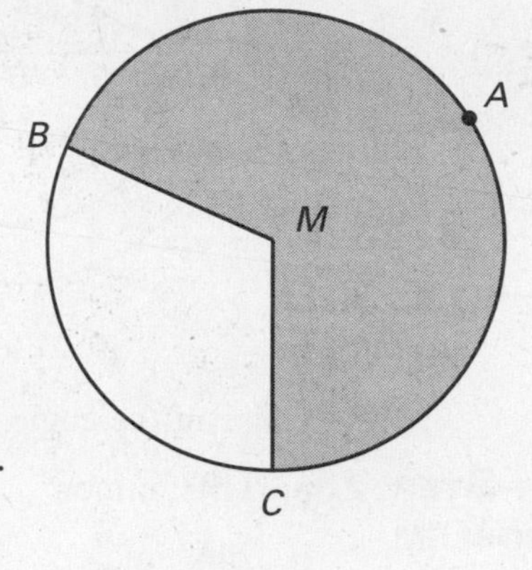

a. Find $m\overset{\frown}{CB}$ = __?__. $\overset{\frown}{CB}$ is a __?__ arc.

b. Find $m\overset{\frown}{CAB}$ = __?__. $\overset{\frown}{CAB}$ is a __?__ arc.

c. Draw $\angle BAC$. Find $m\angle BAC$.

d. The distance between you and the ladder is 12 feet. Write an equation to model the outside of the pool. Assume you are standing at the origin.

e. Write an equation to model the outside of the pool if you are standing at the point $(5, -7)$.

3. ***Critical Thinking*** Use the diagram from Exercise 2.

a. Everyone is out of the pool. The basketball is the only thing left in the pool. The basketball is floating at a point D. $\overline{WY}$ and $\overline{XZ}$ are chords that intersect at D. $WD = 4$, $YD = 5$, and $XD = 2$. Find the length of $\overline{ZD}$.

b. You are standing outside of the pool. You form a tangent segment with the ladder and a secant segment with the basketball hoop. You are 12 feet from the ladder and 6 feet from the edge of the pool along a direct path to the basketball hoop. How far are you from the basketball hoop?

4. ***Writing*** When two lines intersect a circle, there are three places in relation to the circle where the lines intersect each other. Represent each case with a diagram and the appropriate labels. Explain how to find the angle measure for each case.

Alternative Assessment Rubric

For use after Chapter 10

JOURNAL SOLUTION

1. Complete answers should include:

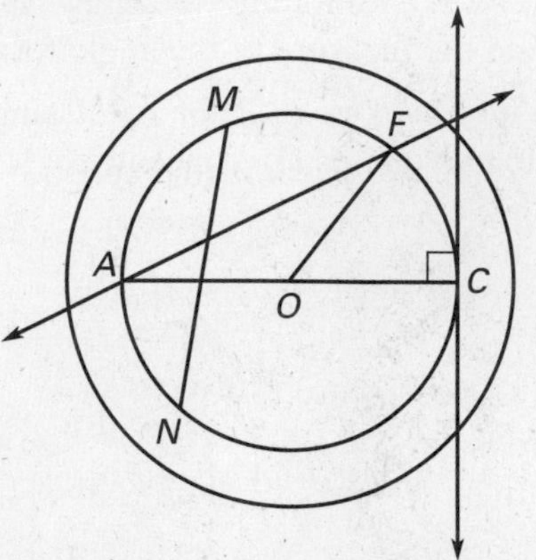

Measure of the angle formed by the tangent and $\overline{AC} = 90°$

MULTI-STEP PROBLEM SOLUTION

2. **a.** 114°; minor

   b. 246°; major

   c. 57°

   d. $x^2 + y^2 = 144$

   e. $(x - 5)^2 + (y + 7)^2 = 144$

3. **a.** 10

   b. 24 feet

4. Answers should include the three following cases: on the circle, inside the circle, and outside the circle.

MULTI-STEP PROBLEM RUBRIC

4 Students answer all parts of the problem correctly, showing work in a step-by-step manner. Students explain how to find the angle measures correctly for all three cases. Students have all diagrams correctly labeled.

3 Students complete the questions. Students' work may have minor mathematical errors. Students explain how to find the angle measures for all three cases. Students may have an incorrect label in the diagrams.

2 Students complete the questions. Students' work may contain several mathematical errors. Students' explanation on how to find angle measures is insufficient. Students have diagrams with incorrect labels.

1 Students do not complete the questions. Students' explanations on how to find angle measures are incomplete. Students' diagrams do not match the explanations.

ANSWERS

Chapter Support

Parent Guide

10.1: no; $\triangle ABC$ is not a right triangle because $5^2 + 12^2 \neq 14^2$. **10.2:** 80°; 280°; 140°; 140°

10.3: 40° **10.4:** about 8° **10.5:** 6 ft

10.6: yes; the point $(20, -20)$ is inside the listening area which is defined by the equation $x^2 + y^2 = 30^2$. **10.7:** at $(-2, -2)$

Prerequisite Skills Review

1. 5, −5 **2.** 3, −3 **3.** −9 **4.** 5, −5 **5.** (6, 4) **6.** (−1, 2) **7.** $YZ = 6, m\angle X = 37°$, $m\angle Z = 53°$ **8.** $JK = 8.0, JL = 6.9$, $m\angle K = 60°$ **9.** $MN = 6.4, NO = 7.7$, $m\angle M = 50°$ **10.** $WY = 8.5, m\angle W = 69°$, $m\angle Y = 21°$ **11.** $PQ = 7.7, QR = 2.2$, $m\angle P = 15°$ **12.** $DE = 12, m\angle E = 23°$, $m\angle C = 67°$ **13.** 6.3, (1, 6) **14.** 8.2, (4, 1) **15.** 10.2, (−1, 3) **16.** 13, $\left(6, \frac{5}{2}\right)$ **17.** 11.2, $\left(-\frac{3}{2}, 2\right)$ **18.** 17, $\left(1, \frac{3}{2}\right)$

Strategies for Reading Mathematics

1. *Sample answer:* To provide a short way to identify the postulate or theorem. **2.** *Sample answer:* There may be a diagram on the page given that can help you understand the postulate or theorem. **3.** *Sample answer:* If a postulate or theorem is used often, we give it a name so that we can refer to it easily, without stating the entire postulate or theorem. **4.** *Sample answer:* So you have a reference of all the postulates and theorems to use when you write proofs or solve problems.

Lesson 10.1

Warm-Up Exercises

1. 5 **2.** 6 **3.** 6; −6 **4.** $4\sqrt{3}; -4\sqrt{3}$ **5.** 3; −7

Daily Homework Quiz

1. $\langle 3, 7\rangle$; 7.6 **2.** $\langle 7, -6\rangle$; 9.2 **3.** $\langle 4, 3\rangle$ **4.** $\langle 11, 0\rangle$

Lesson Opener

Allow 5 minutes.

Note: An easy-to-understand demonstration of how a steam engine works can be found on the Internet at www.howstuffworks.com.

1. *Sample answers:*

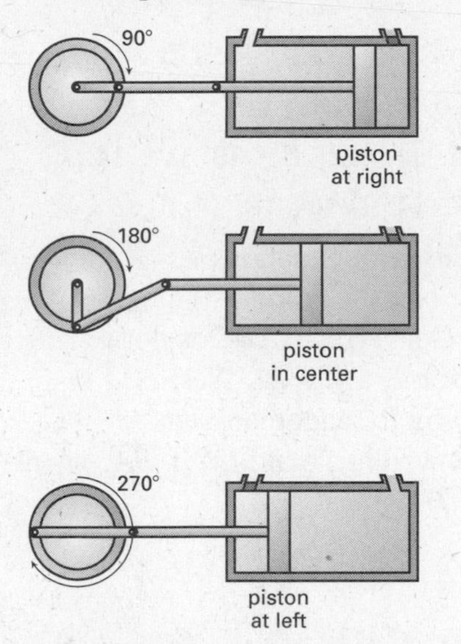

2. The throw of the crank is half the distance the piston moves. (Note: You can figure this out from the diagrams in Ex.1. When piston is at left, distance from the center of flywheel to piston equals length of connecting rod + length of piston rod − throw. When piston is at right, distance from center of flywheel to piston equals length of connecting rod + length of piston rod + throw. Therefore, as the piston moves from left to right, it covers a distance equal to twice the throw.) *Sample answer:* If the throw is 5 inches long, then the cylinder would have to be 10 inches.

Technology Activity

1. Check student's work.

Practice A

1. 3 in. **2.** 12 cm **3.** 7.5 ft **4.** 4.5 in. **5.** 22 cm **6.** 16 ft **7.** 20 in. **8.** 9.2 cm **9.** E **10.** G **11.** D **12.** B **13.** A **14.** C

Lesson 10.1 *continued*

15. H **16.** F **17.** (3, 2), 2 **18.** (7, 2), 2 **19.** (5, 2) **20.** lines with equations $y = 0$, $y = 4$, $x = 5$ **21.** Yes; they have equal radii. **22.** Yes; $5^2 + 12^2 = 13^2$ so by Pythagorean Thm. Converse $\triangle ABC$ is a right $\triangle$ and $\overline{CA} \perp \overline{BA}$. Then $\overleftrightarrow{AB}$ is a tangent to $\odot C$. **23.** No; $8^2 + 16^2 \neq 20^2$ so by Pythagorean Thm. Converse, $\triangle ABC$ is not a right $\triangle$. So $\overline{AB}$ is not $\perp$ $\overline{CA}$ so $\overleftrightarrow{AB}$ is not a tangent. **24.** 330 ft, 660 ft

Practice B

1. 6.5 in. **2.** 4 cm **3.** 6.3 ft **4.** 1 ft $2\frac{1}{2}$ in. **5.** 34 cm **6.** 12.6 ft **7.** 1.5 in. **8.** 8.5 ft **9.** E **10.** G **11.** D **12.** B **13.** A **14.** C **15.** H **16.** F **17.** (4, 2), 2 **18.** (4, 6), 2 **19.** (4, 4) **20.** lines with equations $y = 4$, $x = 2$, $x = 6$ **21.** No; $8^2 + 12^2 \neq 17^2$ so by Pythagorean Thm. Converse, $\triangle ABC$ is not a right $\triangle$. So $\overline{AB}$ is not $\perp$ $\overline{CA}$. **22.** Yes; $7^2 + 24^2 = 25^2$ so by Pythagorean Thm. Converse $\triangle ABC$ is a right $\triangle$ and $\overline{CA} \perp \overline{BA}$. Then $\overleftrightarrow{AB}$ is tangent to $\odot C$.

23. 2.5 **24.** −5, 5 **25.** −2, 2

Practice C

1. point of tangency
2. common internal tangent **3.** radius
4. chord **5.** center **6.** diameter **7.** secant
8. common external tangent

9. **10.** **11.**

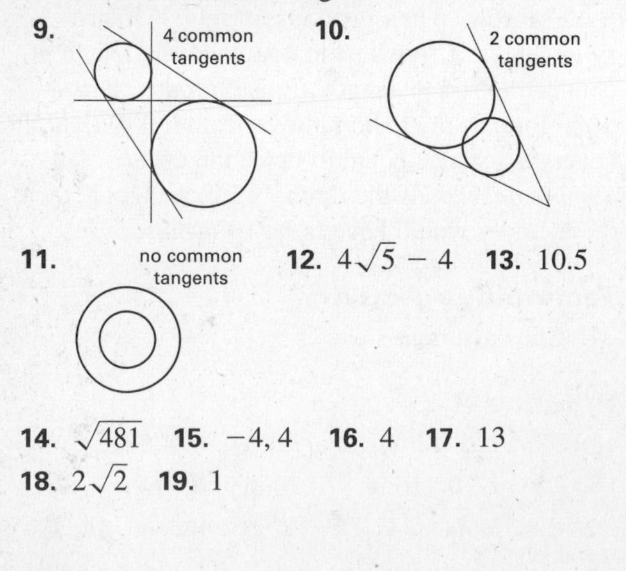

12. $4\sqrt{5} - 4$ **13.** 10.5

14. $\sqrt{481}$ **15.** −4, 4 **16.** 4 **17.** 13 **18.** $2\sqrt{2}$ **19.** 1

Reteaching with Practice

1. chord **2.** secant **3.** radius **4.** diameter **5.** tangent **6.** radius **7.** chord **8.** radius **9.** −5, 5 **10.** 4 **11.** 16

Real-Life Application

1. B and C **2.** $\overline{DB}$ and $\overline{DC}$ **3.** Theorem 10.1 **4.** *Sample answer:* Since the triangles are right triangles with $\overline{AD}$ common to both, and $\overline{DC}$ and $\overline{DB}$ are radii, the triangles can be proven congruent with the HL Theorem. **5.** Theorem 10.3

Challenge: Skills and Applications

1. 30
2. *Sample Answer:*

Statements	Reasons
1. $\overline{AC}$ and $\overline{BC}$ are radii of the circle.	**1.** Definition of radius
2. $\overline{BC} \cong \overline{AC}$	**2.** All radii of a circle are congruent.
3. $\overline{AC} \cong \overline{AD}$	**3.** Given
4. $\overline{BC} \cong \overline{AD}$	**4.** Transitive Property of Congruence
5. $\overline{AD}$ is tangent to the circle.	**5.** Given
6. $\overline{AC} \perp \overline{AD}$	**6.** A tangent line is $\perp$ to the radius drawn to the point of tangency.
7. $\overline{AC} \perp \overline{BC}$	**7.** Given
8. $\overline{BC} \parallel \overline{AD}$	**8.** In a plane, 2 lines $\perp$ to the same line are $\parallel$.
9. $ABCD$ is a parallelogram.	**9.** If one pair of opp. sides of a quad. are $\cong$ and $\parallel$, then the quad. is a $\square$.

3. $\odot X$: 12, $\odot Y$: 6, $\odot Z$: 8 **4.** 25
5. a. *Sample answer:* Since a tangent line is perpendicular to the radius drawn to the point of tangency, $\overline{QR} \perp \overline{RS}$ and $\overline{RS} \perp \overline{SP}$. Since $\overline{QT} \perp \overline{SP}$ is given, $\overline{QT} \parallel \overline{RS}$ (in a plane, 2 lines perpendicular to the same line are parallel), which implies $\overline{QT} \perp \overline{QR}$ (if a transversal is perpendicular to one of two parallel lines, then it is perpendicular to the other). So, $QRST$ has four right angles; by the Rectangle Corollary, $QRST$ is a rectangle. **b.** $6\sqrt{3}$ in. **c.** 60°

Lesson 10.2

Warm-Up Exercises

1. 25 **2.** 49 **3.** 12 **4.** 3; −3 **5.** 30

Daily Homework Quiz

1. 26 ft **2.** 6.4 in. **3.** yes **4.** 7

Lesson Opener

Allow 15 minutes.

1. Any point on the perpendicular bisector of $\overline{AB}$ is equidistant from A and B. This is necessary because the circle must pass through both A and B. **2.** *Sample answer:* $C(9, 1)$, $AC = \sqrt{50} \approx 7.1$ units, $m\angle ACB \approx 127°$. There are an infinite number of choices for the location of C, since C can be anywhere on the perpendicular bisector of $\overline{AB}$.

3. *Sample answer:*

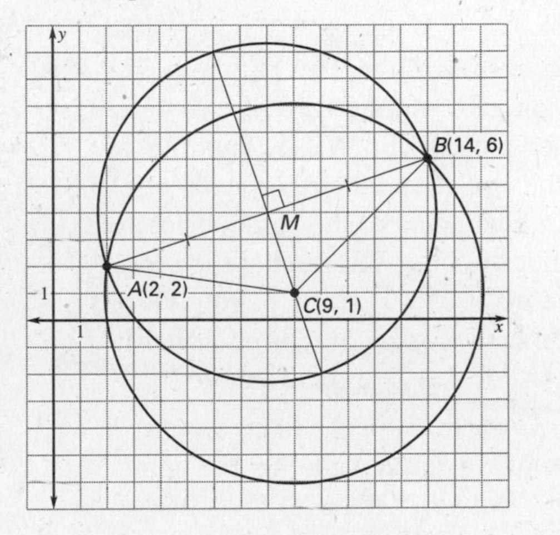

No. In most situations there would be no need for such a big curve to get from point A to point B. Also, this arc is likely to result in dangerously sharp turns in the road at points A and B. (An exceptional case in which this arc might be suitable is if the road is supposed to form a loop so that, after going around the curve, you are going back in the opposite direction. That is, the sections of the road approaching the curve at points A and B are almost parallel.) **4.** The arc gets shorter. The angle measure gets smaller.

Practice A

1. minor arc **2.** semicircle **3.** semicircle **4.** major arc **5.** minor arc **6.** minor arc **7.** semicircle **8.** minor arc **9.** 70° **10.** 110° **11.** 180° **12.** 210° **13.** 70° **14.** 110° **15.** 290° **16.** 30° **17.** 280° **18.** 290° **19.** 130° **20.** 165° **21.** 60° **22.** $\overset{\frown}{AB} \cong \overset{\frown}{DC}$; if two chords of the same ⊙ are ≅, then their minor arcs are ≅. **23.** $\overline{AB} \cong \overline{CB}$; if two minor arcs of the same ⊙ are ≅, then their chords are ≅. **24.** $\overline{AD} \cong \overline{BD}$; if a diameter of a ⊙ is ⊥ to a chord, then it bisects the chord. **25.** 8 **26.** 8 **27.** 12

Practice B

1. minor arc **2.** semicircle **3.** semicircle **4.** major arc **5.** minor arc **6.** minor arc **7.** semicircle **8.** minor arc **9.** 63° **10.** 117° **11.** 180° **12.** 215° **13.** 63° **14.** 117° **15.** 297° **16.** 35° **17.** 278° **18.** 297° **19.** 90° **20.** 105° **21.** 100° **22.** C **23.** D **24.** A **25.** E **26.** B **27.** F **28.** 4 **29.** 80°

Practice C

1. 73° **2.** 107° **3.** 180° **4.** 206° **5.** 81° **6.** 180° **7.** 73° **8.** 107° **9.** 287° **10.** 26° **11.** 279° **12.** 287° **13.** 106° **14.** 66° **15.** 48° **16.** 3; $\overline{PX} \cong \overline{PY} \cong \overline{PZ}$ since they are radii of ⊙P; $\angle XPY \cong \angle ZPY$; $\triangle XPY \cong \triangle ZPY$ by SAS Congruence Postulate, so $\overline{XY} \cong \overline{ZY}$ because corresponding parts of ≅ △s are ≅. **17.** 5; diameter $\overline{XV}$ bisects $\overline{ZY}$ so $YW = 3$; $\triangle XWY$ is a 3-4-5 right triangle. **18.** 6; $\overline{AB} \cong \overline{XY}$ since they are chords equidistant from the center; since $CA = 3$, $AB = 6$ by knowing C is the midpoint of $\overline{AB}$; then if $AB = 6$, $XY = 6$

Lesson 10.2 *continued*

19.

Statements	Reasons
1. $\overline{AG}$ is diameter of $\odot P$.	**1.** Given
2. $\overline{AG} \perp \overline{CD}$	**2.** Given
3. $\overline{CF} \cong \overline{DF}$	**3.** A diameter $\perp$ to a chord bisects the chord.
4. $\overline{AF} \cong \overline{AF}$	**4.** Reflexive Property of $\cong$
5. $\angle AFC \cong \angle AFD$	**5.** Def of $\perp$ and all rt $\angle$s are $\cong$.
6. $\triangle AFC \cong \triangle AFD$	**6.** SAS Congruence Postulate
7. $\overline{AC} \cong \overline{AD}$	**7.** Corresponding parts of $\cong$ $\triangle$'s are $\cong$.

20.

Statements	Reasons
1. $\overline{CQ} \cong \overline{AP}$	**1.** Given
2. $\odot P \cong \odot Q$	**2.** Def of $\cong$ $\odot$s
3. $\overline{DQ} \cong \overline{BP}$	**3.** Radii of $\cong$ $\odot$s are $\cong$.
4. $\overparen{AB} \cong \overparen{CD}$	**4.** Given
5. $\overline{AB} \cong \overline{CD}$	**5.** $\cong$ arcs of $\cong$ $\odot$s cut off $\cong$ chords.
6. $\triangle APB \cong \triangle CQD$	**6.** SSS Congruence Thm.

Reteaching with Practice

1. a. 180° **b.** 90° **c.** 90° **d.** 270°

2. a. 110° **b** 200° **c.** 160° **d.** 250°

3. $\sqrt{11} \approx 3.3$ **4.** 3

Cooperative Learning Activity

1. They are congruent. **2.** They are congruent.

3. no

Interdisciplinary Application

1. *Sample answer:*

2. diameter **3.** The center point; diameters are chords that pass through the center. Since each of the bisectors are diameters, then any point they have in common must be the center. **4.** Place a compass with the needle on the center point and the pencil tip on some point on the partial circle. Draw the circle.

Challenge: Skills and Applications

1. *Sample answer:* Since three points determine a plane, let P be the plane that contains X, Y, and Z. Let line j be the perpendicular bisector of $\overline{XY}$ in P, and let line k be the perpendicular bisector of $\overline{YZ}$ in P. Note that j and k are not parallel; for, if they were parallel, then $\overleftrightarrow{XY}$ and $\overleftrightarrow{YZ}$ would be either parallel lines or the same line, both of which are impossible or contradict the given information. Since j and k are nonparallel lines in P, they contain a common point C. By the Perpendicular Bisector Theorem, $CX = CY$ and $CY = CZ$, so the circle centered at C with radius CY contains X, Y, and Z.

2. *Sample answer:* Given $\triangle XYZ$, the points X, Y, and Z are noncollinear and therefore (by Exercise 1) they lie on a circle. So, $\overline{XY}$, $\overline{YZ}$, and $\overline{XZ}$ are chords of the circle, and their perpendicular bisectors are diameters of the circle. Therefore, the perpendicular bisectors intersect at the center C of the circle. Since $\overline{CX}$, $\overline{CY}$, and $\overline{CZ}$ are radii of the circle, the intersection point C is equidistant from the vertices X, Y, and Z of the triangle.

3. *Sample answer:* Suppose there are three collinear points on circle C. Label the points X, Y, and Z, with Y between X and Z. Then the perpendicular bisectors of $\overline{XY}$ and $\overline{YZ}$ are parallel lines, but they both contain the center C. This is a contradiction, so there cannot be three distinct, collinear points on a circle.

4. *Sample answer:* Select a point R on the arc. Construct the perpendicular bisectors of $\overline{PR}$ and $\overline{RQ}$; the intersection point of the bisectors is the center C of the circle. Construct a circle with center C and radius CP.

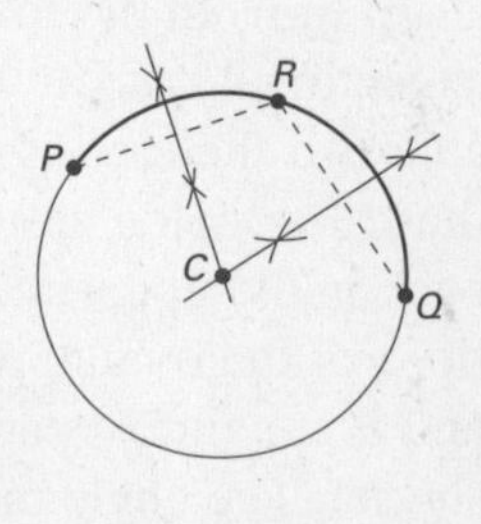

Lesson 10.2 *continued*

5. *Sample answer:* Construct perpendicular bisectors of $\overline{DG}$ and $\overline{EH}$. We may label the bisectors $\overleftrightarrow{CK}$ and $\overleftrightarrow{CL}$, where C is the center of the circle, K is the midpoint of $\overline{DG}$ and L is the midpoint of $\overline{EH}$. Then $\overline{CK} \cong \overline{CL}$, because congruent chords are equidistant from the center, and $\overline{CJ} \cong \overline{CJ}$, by the Reflexive Property of Congruence. Therefore, by the HL Congruence Theorem, $\triangle KJC \cong \triangle LJC$; this gives $\angle DJI \cong \angle HJI$ because corresponding parts of $\cong \triangle$s are $\cong$. **6.** $m\widehat{DE} = 50°$, $m\widehat{EF} = 130°$, $m\widehat{FG} = 50°$, $m\widehat{GD} = 130°$ **7.** 73 **8.** 12 **9.** 13

Lesson 10.3

Warm-Up Exercises

1. 22.5 **2.** 18 **3.** 32 **4.** $\langle 40, 20 \rangle$

Daily Homework Quiz

1. major arc **2.** minor arc **3.** 78° **4.** 220° **5.** 28

Lesson Opener

Allow 15 minutes.

1. *Sample answer:*

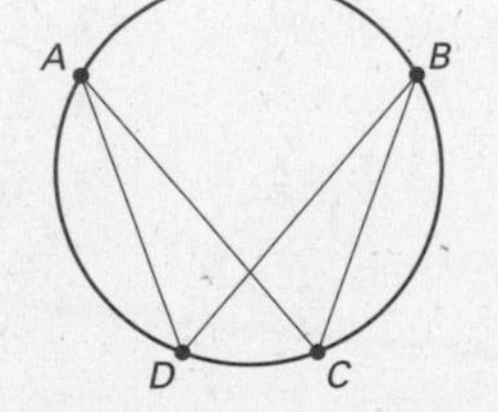

2. $m\angle ADB = m\angle BCA$; The rays that form each angle intersect the circle at A and B, the endpoints of arc AB. **3.** $m\angle DAC = m\angle CBD$; arc CD **4.** $MP = MQ$

Practice A

1. 76° **2.** 156° **3.** 80° **4.** 180° **5.** 12° **6.** 16° **7.** 60° **8.** 110° **9.** 35° **10.** 30° **11.** 130° **12.** 180° **13.** 60° **14.** 250° **15.** no **16.** yes **17.** yes **18.** $x = 180, y = 90$ **19.** $x = 55, y = 55$ **20.** $x = 75$

Practice B

1. 66° **2.** 54° **3.** 43° **4.** 90° **5.** 54° **6.** 50° **7.** 54° **8.** 36° **9.** 50° **10.** 72° **11.** 180° **12.** 23.5 **13.** 7 **14.** 102 **15.** 13 **16.** 23.25 **17.** 8

18. *Sample answer:* Draw a chord, construct ⊥ bisector; draw a second chord and construct ⊥ bisector. Where ⊥ bisectors intersect is the center. Measure radius. Double for diameter.

Practice C

1. 74° **2.** 132° **3.** 43.5° **4.** 56° **5.** 19° **6.** 21° **7.** 90° **8.** 43° **9.** 47.5° **10.** 43° **11.** 47° **12.** 47.5° **13.** 94° **14.** 180° **15.** 20 **16.** 7 **17.** 36

18.

Statements	Reasons
1. $\angle MEI \cong \angle GED$	1. Vert. Angles Thm.
2. $m\angle IMD = \frac{1}{2} m\widehat{ID}$	2. Measure of inscribed $\angle = \frac{1}{2}$ measure of intercepted arc.
3. $m\angle IGD = \frac{1}{2} m\widehat{ID}$	3. Measure of inscribed $\angle = \frac{1}{2}$ measure of intercepted arc.
4. $m\angle IMD = m\angle IGD$	4. Trans. Prop. of $\cong$
5. $\angle IMD \cong \angle IGD$	5. Def. of $\cong$ $\angle$s
6. $\triangle MEI \sim \triangle GED$	6. AA Similarity Postulate

Reteaching with Practice

1. 34 **2.** 23 **3.** 43 **4.** 50 **5.** 25 **6.** 11 **7.** $x = 40, y = 93$ **8.** $x = 56, y = 20$ **9.** $x = 30, y \approx 59.3$

Real-Life Application

1. $\angle BDC$ **2.** 4400 miles **3.** 25° **4.** 50° **5.** 8 **6.** about 3666 miles

Challenge: Skills and Applications

1. *Sample answer:* Draw $\overline{DG}$. Since $\overline{DF}$ is a diameter, $\angle DGF$ is a right angle inscribed in $\odot C$; therefore, $\overline{DG} \perp \overline{FG}$. So, we have $\overline{FG} \cong \overline{GE}$ (given), $\angle DGF \cong \angle DGE$ (all right angles are

Lesson 10.3 *continued*

congruent), and $\overline{DG} \cong \overline{DG}$ (Reflexive Property of Congruence). By the SAS Congruence Postulate, $\triangle DGF \cong \triangle DGE$, so $\overline{DF} \cong \overline{DE}$ (corresponding parts of $\cong$ $\triangle$s are $\cong$.) and $\triangle DEF$ is isosceles.

2. *Sample answer:* Draw $\overline{PR}$, $\overline{PS}$, and $\overline{PT}$. Since $\overline{PR}$ is a diameter of $\odot Q$, $\angle PSR$ is a right angle inscribed in $\odot Q$; therefore, $\overline{PS} \perp \overline{RT}$. $\triangle PSR$ and $\triangle PST$ are right triangles, $\overline{PR} \cong \overline{PT}$ (radii of a circle are congruent), and $\overline{PS} \cong \overline{PS}$ (Reflexive Property of Congruence). Therefore, $\triangle PSR \cong \triangle PST$ by HL Congruence Theorem, and so $\overline{RS} \cong \overline{ST}$ (corresponding parts of $\cong$ $\triangle$s are $\cong$).

3. *Sample answer:* Draw $\overline{XZ}$. Note that $\overline{WZ} \parallel \overline{XY}$, so by the Alternate Interior Angles Theorem, $\angle WZX \cong \angle YXZ$. But $m\widehat{WX} = 2m\angle WZX$ and $m\widehat{YZ} = 2m\angle YXZ$, so $m\widehat{WX} \cong \widehat{YZ}$. Since two minor arcs in the same circle are congruent if and only if their corresponding chords are congruent, we conclude that $\overline{WX} \cong \overline{YZ}$, so $WXYZ$ is an isosceles trapezoid.

4. a. $OR = c$, $PS = c - a$ **b.** right triangle

c. *Sample answer:* $\dfrac{PS}{PQ} = \dfrac{PQ}{RP}$

d. *Sample answer:* $\dfrac{c-a}{b} = \dfrac{b}{c+a}$,

$b^2 = (c - a)(c + a) = c^2 - a^2$, so $a^2 + b^2 = c^2$.

Quiz 1

1. 90°; a tangent line is $\perp$ to the radius drawn to the point of tangency. **2.** 15; two tangent segments with the same exterior endpoint are congruent. **3.** 60° **4.** 60° **5.** 180° **6.** 240° **7.** 120° **8.** 300° **9.** 102.8°

Lesson 10.4

Warm-Up Exercises

1. 45 **2.** 4 **3.** 5 **4.** 60

Daily Homework Quiz

1. $a = 90°$, $b = 58°$, $c = 45°$ **2.** $x = 6$, $y = 9$

Lesson Opener

Allow 10 minutes.

1. a. A **b.** D, E, F **c.** B, C **2. a.** A, B, F, **b.** E **c.** C, D **3. a.** C, E **b.** B, D **c.** A, F

Practice A

1. 90° **2.** 86° **3.** 116° **4.** 39° **5.** 82° **6.** 74° **7.** 40° **8.** 24.5° **9.** 36°

10. $38° = \frac{1}{2}(180° - x°)$; 104

11. $115° = \frac{1}{2}(105° + x°)$; 125

12. $96° = \frac{1}{2}(360° - x°)$; 168

13. $m\angle 3, m\angle 2, m\angle 1$ **14.** $m\angle 1, m\angle 2, m\angle 3$

Practice B

1. 63° **2.** 148° **3.** 33° **4.** 49° **5.** 13.5°

6. 48° **7.** $104° = \frac{1}{2}(360° - x°)$; 152

8. $22° = \frac{1}{2}(125 - x°)$; 81

9. $38° = \frac{1}{2}(x° + 69°)$; 7

10. $45° = \frac{1}{2}(360 - x° - x°)$; 135

11. $17° = \frac{1}{2}(x° - 42°)$; 76

12. $138° = \frac{1}{2}(360 - x)°$; 84 **13.** $\approx 6.3°$

Practice C

1. 142° **2.** 65° **3.** 40° **4.** 28° **5.** 57.5°

6. 52° **7.** 20° **8.** 100°

9.

Statements	Reasons
1. C is midpt. of $\widehat{BD}$	**1.** Given
2. $\widehat{BC} \cong \widehat{CD}$	**2.** Def. of midpoint
3. $m\angle BAC = \frac{1}{2}\, m\widehat{BC}$ $m\angle CAD = \frac{1}{2}\, m\widehat{CD}$	**3.** In a $\odot$, measure of inscribed $\angle = \frac{1}{2}$ measure intercepted arc
4. $m\widehat{BC} = m\widehat{CD}$	**4.** Congruent arcs have = measure
5. $\frac{1}{2}\, m\widehat{BC} = \frac{1}{2}\, m\widehat{CD}$	**5.** Mult. Prop. of Equality
6. $m\angle BAC = m\angle CAD$	**6.** Substitution
7. $\overrightarrow{AC}$ bisects $\angle BAD$	**7.** Def of $\angle$ bisector

Lesson 10.4 *continued*

10.

Statements	Reasons
1. $\overline{BC} \cong \overline{CD}$	**1.** Given
2. Circle with center E	**2.** Given
3. $\widehat{BC} \cong \widehat{CD}$	**3.** Congruent chords in the same ⊙ cut off ≅ arcs
4. $m\angle BAC = \frac{1}{2} m\widehat{BC}$ $m\angle DAC = \frac{1}{2} m\widehat{CD}$	**4.** An inscribed ∠ of a ⊙ = $\frac{1}{2}$ measure of intercepted arc
5. $m\widehat{BC} = m\widehat{CD}$	**5.** ≅ arcs have = measure
6. $\frac{1}{2} m\widehat{BC} = \frac{1}{2} m\widehat{CD}$	**6.** Mult. Prop. of Equality
7. $m\angle BAC = m\angle DAC$	**7.** Substitution
8. $\angle BAC \cong \angle DAC$	**8.** Def. of ≅ ∠s
9. $\angle B$ is a rt. ∠ $\angle D$ is a rt. ∠	**9.** An ∠ inscribed in a semicircle is a rt. ∠
10. $\angle B \cong \angle D$	**10.** All rt. ∠s are ≅
11. $\triangle ABC \cong \triangle ADC$	**11.** AAS Congruence Postulate

Reteaching with Practice

1. $x = 78, y = 102$ **2.** $x = 180$
3. $x = 60, y = 240, z = 120$
4. 65 **5.** 70 **6.** 30 **7.** 49 **8.** 70 **9.** 20

Interdisciplinary Application

1. 90° **2.** 180° **3.** 13° **4.** 6.5° **5.** 6.5°

Challenge: Skills and Applications

1. *Sample answer:* Draw line m tangent to both circles at T. Let J be the point where line k intersects line m, and let K be the point where line j intersects line m. Now $m\widehat{TU} = m\widehat{TV}$ (because both are equal to $2\angle KTU$) and $m\widehat{TWU} = m\widehat{TXV}$ (because both are equal to $360° - 2m\angle KTU$). Since $m\angle TJU = \frac{1}{2}(m\widehat{TWU} - m\widehat{TU})$ and $m\angle TKV = \frac{1}{2}(m\widehat{TXV} - m\widehat{TV})$, the substitution property of equality gives $m\angle TJU = m\angle TKV$. Therefore, by the Corresponding Angles Postulate, $j \parallel k$.

2. *Sample answer:* Observe that $m\angle D = \frac{1}{2}(m\widehat{AG} - m\widehat{CE}) = \frac{1}{2}(2m\widehat{CE} - m\widehat{CE}) = \frac{1}{2}m\widehat{CE}$, and $m\angle D = \frac{1}{2}(m\widehat{BHF} - m\widehat{BF})$ $= \frac{1}{2}(360° - m\widehat{BF} - m\widehat{BF}) = 180° - m\widehat{BF}$. Therefore, $\frac{1}{2} m\widehat{CE} = 180° - m\widehat{BF}$, which gives $m\widehat{CE} + 2m\widehat{BF} = 360°$.

3. $(x - y)°$ **4.** 90 **5.** 108 **6.** 12 **7.** 8
8. 10 **9.** 7, 8

Lesson 10.5

Warm-Up Exercises

1. 6 **2.** 18 **3.** $4\frac{1}{3}$ **4.** $-3 \pm 3\sqrt{17}$

Daily Homework Quiz

1. 11 **2.** 5

Lesson Opener

Allow 10 minutes.

1. Answers will vary. **2.** Answers will vary.
3. The products are equal.
4. *Sample answer:*
Conjecture: If two chords intersect inside a circle, then the product of the lengths of the segments of one chord is equal to the product of the lengths of the segments of the other chord.

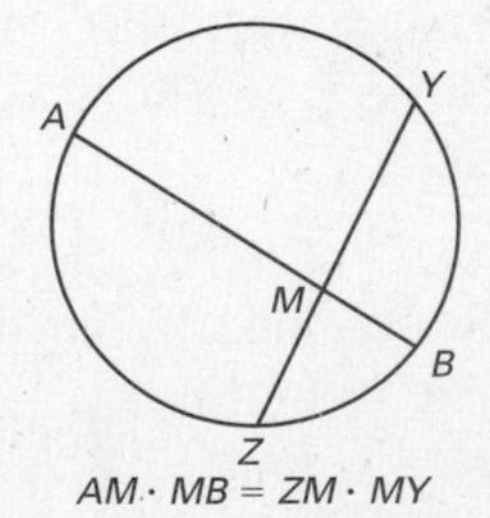

Practice A

1. 3, 9; 15 **2.** 4, x; 8 **3.** 16; 8 **4.** 18; 6
5. x, 15; 25 **6.** x, 12; 16 **7.** 12 **8.** 4
9. $\frac{35}{3}$ **10.** 12 **11.** 9 **12.** 9 **13.** 5.3 feet

Practice B

1. 4, 6; 12 **2.** x, 18; 22.5 **3.** 10; $2\sqrt{15}$
4. 7.4 **5.** 12.5 **6.** 5.7 **7.** 1.9 **8.** 28.1
9. 7.7 **10.** 5.6 **11.** 3 **12.** 2
13. a. 50 cm **b.** 80 cm, 80 cm **c.** 128 cm
d. 178 cm **e.** 89 cm

Lesson 10.5 *continued*

Practice C

1. 3.7 **2.** 2.3 **3.** 7.4 **4.** 3 **5.** 1 **6.** 3.9 **7.** 14.3 **8.** 5 **9.** 10 **10.** 18, 4

11. *Sample answer:* When you use the theorem to solve for x and y you get $x = 26$ and $y = 39$. When the figure is drawn to scale, the segments do not intersect in the interior of both circles, so Theorem 10.15 cannot be applied. **12.** 900 miles

Reteaching with Practice

1. 3.2 **2.** 4 **3.** 6 **4.** 7.25 **5.** ≈ 3.7 **6.** 2 **7.** $\sqrt{55} \approx 7.4$ **8.** $8\sqrt{3} \approx 13.9$ **9.** $\sqrt{\frac{117}{4}} \approx 5.4$

Real-Life Application

1. tangent line **2.** 139,000 km **3.** 238,000 km

Challenge: Skills and Applications

1. *Sample answer:* Since $OP \cdot OQ$ and $OR \cdot OS$ are both equal to $(OT)^2$, $OP \cdot OQ = OR \cdot OS$.

2. a. *Sample answer:* Since vertical angles are congruent, $\angle BEC \cong \angle DEA$ and $\angle CEA \cong \angle BED$. Since $EA \cdot EB = EC \cdot ED$, we may write $\frac{EA}{ED} = \frac{EC}{EB}$, which gives $\triangle CEA \sim \triangle BED$ by the SAS Similarity Theorem. We may also write $\frac{EA}{EC} = \frac{ED}{EB}$, which gives $\triangle DEA \sim \triangle BEC$.

b. *Sample answer:* First note that since $\triangle CEA \sim \triangle BED$, $\angle 4 \cong \angle 7$; since $\triangle DEA \sim \triangle BEC$, $\angle 5 \cong \angle 2$. Therefore,
$m\angle ACB + m\angle ADB$
$= m\angle ACB + (m\angle 5 + m\angle 4)$
$= m\angle ACB + m\angle 2 + m\angle 7 = 180°$, since the sum of the measures of the angles of $\triangle ABC$ is 180°. **c.** *Sample answer:* We have shown that $\angle ACB$ and $\angle ADB$ are supplementary. By similar reasoning, $\angle CAD$ and $\angle CBD$ are supplementary. Therefore, $ACBD$ is a quadrilateral whose opposite angles are supplementary, which implies that $ACBD$ can be inscribed in a circle. (A quadrilateral can be inscribed in a circle if and only if its opposite angles are supplementary.) So, there exists a circle containing A, B, C, and D.

3. a. 90° **b.** $\frac{OP}{OQ} = \frac{OQ}{OR}$ **c.** *Sample answer:* The theorem involving segments of chords gives $OP \cdot OR = OS \cdot OQ$. But $OS = OQ$, so $OP \cdot OR = (OQ)^2$, or $\frac{OP}{OQ} = \frac{OQ}{OR}$.

4. 3 **5.** 6 **6.** 5

Quiz 2

1. 25 **2.** 60 **3.** 240 **4.** $2\frac{3}{4}$ **5.** 12.5 **6.** 4

7. Solve $18(18 + 2r) = 36^2$ (Thm. 10.17) or solve $(r + 18)^2 = r^2 + 36^2$ (Pythagorean Thm.); 27 ft

Lesson 10.6

Warm-Up Exercises

1. 12 **2.** 5 **3.** $\sqrt{26}$ **4.** $4\sqrt{2}$ **5.** $\sqrt{29}$

Daily Homework Quiz

1. 6 **2.** 8

Lesson Opener

Allow 15 minutes.

1. Check art: circle with center (0, 0) and radius 5; $x^2 + y^2 = 25$ **2.** Check art: circle with center (0, 0) and radius 9 **3.** Check art: circle with center (2, −2) and radius 2; $(x - 2)^2 + (y + 2)^2 = 4$ **4.** Check art: circle with center (−5, 3) and radius 3

Technology Activity

1. Given a circle with equation $(x - a)^2 + (y - b)^2 = r^2$, the center is (a, b) and the radius is r. **2.** $(-b, c)$ and $\sqrt{d}$

3. $5 + \sqrt{15} \approx 8.87$, $5 - \sqrt{15} \approx 1.13$

4. a. outside **b.** inside **c.** outside **d.** on

Practice A

1. C **2.** F **3.** A **4.** B **5.** E **6.** D

7. (0, 0), 5 **8.** (0, 4), 3 **9.** (5, 0), 4

10. (−1, 1), 2 **11.** (2, 4), 4 **12.** (−4, 2), 5

13. (0, 0), 3; $x^2 + y^2 = 9$

14. (0, 2), 2; $x^2 + (y - 2)^2 = 4$

15. (1, 1), 1; $(x - 1)^2 + (y - 1)^2 = 1$

16. $x^2 + y^2 = 4$ **17.** $x^2 + (y - 1)^2 = 4$

Lesson 10.6 *continued*

18. $(x - 2)^2 + y^2 = 9$

19. $(x - 3)^2 + (y - 3)^2 = 16$

20. interior **21.** exterior **22.** on **23.** on

24. exterior **25.** interior

Practice B

1. C **2.** F **3.** A **4.** B **5.** E **6.** D

7. $(4, -2), 5$ **8.** $(-2, -4), 3$ **9.** $(5, 3), 4$

10. $(-6, 4), 2$ **11.** $(5, 6), 6$ **12.** $(-3, 4), 4$

13. $(-2, 0), 4; (x + 2)^2 + y^2 = 16$

14. $(0, -3), 3; x^2 + (y + 3)^2 = 9$

15. $(3, 3), 3; (x - 3)^2 + (y - 3)^2 = 9$

16. $x^2 + y^2 = 1$ **17.** $x^2 + (y - 4)^2 = 16$

18. $(x + 4)^2 + (y - 2)^2 = 9$

19. $(x + 3)^2 + (y + 5)^2 = 25$

20. interior **21.** exterior **22.** on **23.** on

24. exterior **25.** exterior

Practice C

1. C **2.** F **3.** A **4.** B **5.** E **6.** D

7. $(3, -5), 6$ **8.** $(-4, -2), 9$ **9.** $(9, 5), 2\sqrt{10}$

10. $(-1.5, 3.8), 1.2$ **11.** $\left(\frac{1}{2}, \frac{3}{4}\right), \frac{2}{3}$

12. $\left(-\frac{3}{5}, \frac{1}{10}\right), \frac{3}{5}$

13. $(2, 2), 4; (x - 2)^2 + (y - 2)^2 = 16$

14. $(-3, 2), 2; (x + 3)^2 + (y - 2)^2 = 4$

15. $(5, -2), 3; (x - 5)^2 + (y + 2)^2 = 9$

16. $x^2 + (y - 4)^2 = 25$

17. $(x + 3)^2 + (y - 6)^2 = 49$

18. $(x - 4.2)^2 + (y - 2.6)^2 = 12.25$

19. $\left(x - \frac{7}{2}\right)^2 + \left(y - \frac{5}{2}\right)^2 = 4$

20.

21.

22.

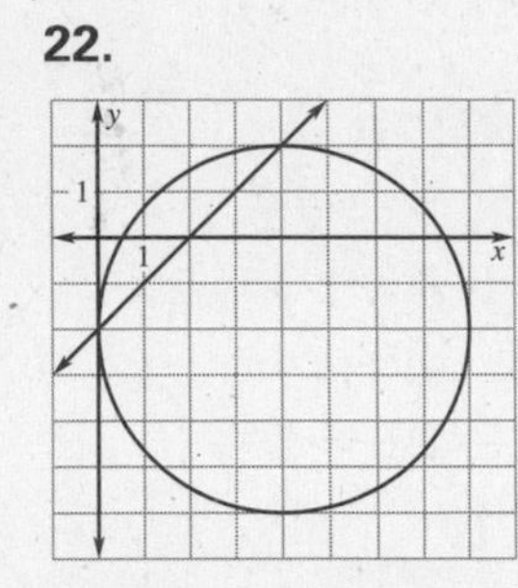

secant; line intersects ⊙ twice

23.

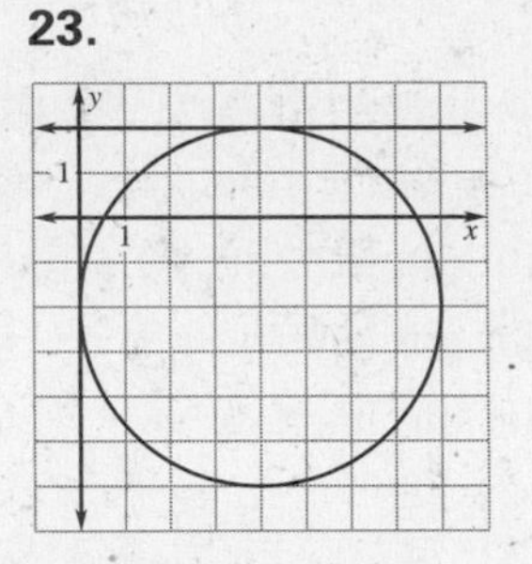

tangent; line intersects ⊙ once

24.

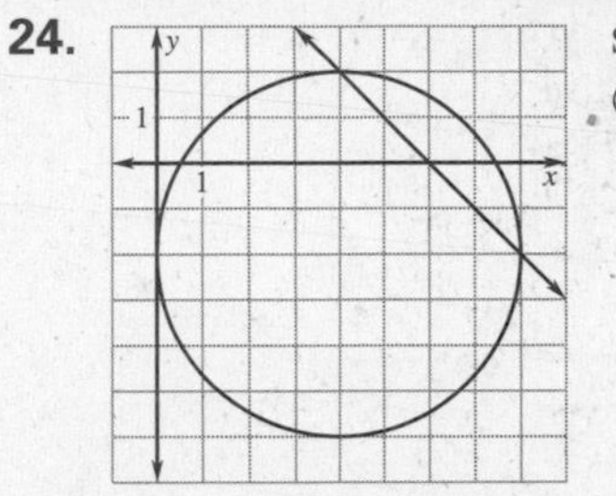

secant; line intersects ⊙ twice

Reteaching with Practice

1. $(x - 4)^2 + (y + 1)^2 = 36$

2. $(x + 1)^2 + (y + 5)^2 = 10.24$

3. $(x + 2)^2 + (y - 3)^2 = 16$

4.

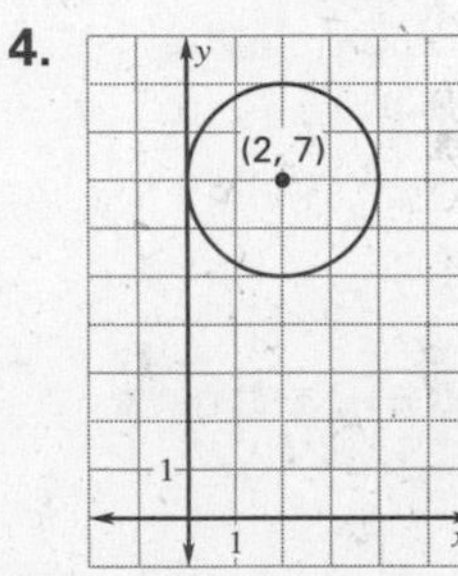

5.

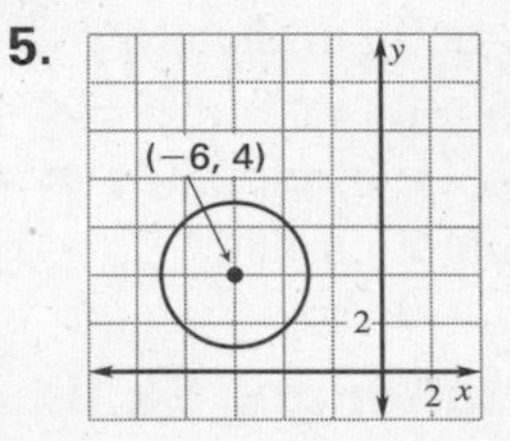

6.

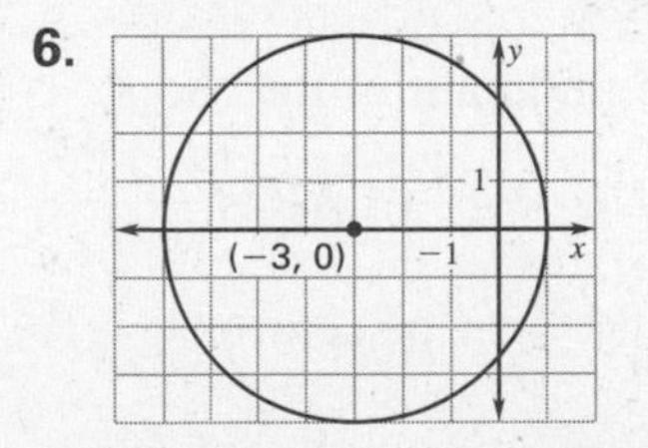

7.

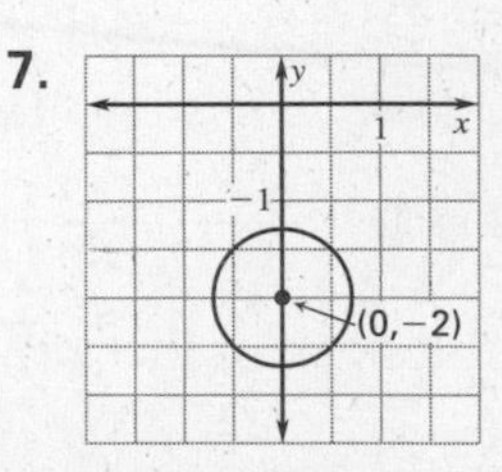

8. no **9.** yes **10.** no

Cooperative Learning Activity

1. $x^2 + y^2 = 30^2$ **2.** Plane 3 **3.** radius

Lesson 10.6 *continued*

Interdisciplinary Application

1. 12-point circle: $(x-8)^2+(y-4)^2=2^2$
9-point circle: $(x-8)^2+(y-4)^2=3^2$
3-point circle: $(x-8)^2+(y-4)^2=4^2$
1-point circle: $(x-8)^2+(y-4)^2=5^2$

2.

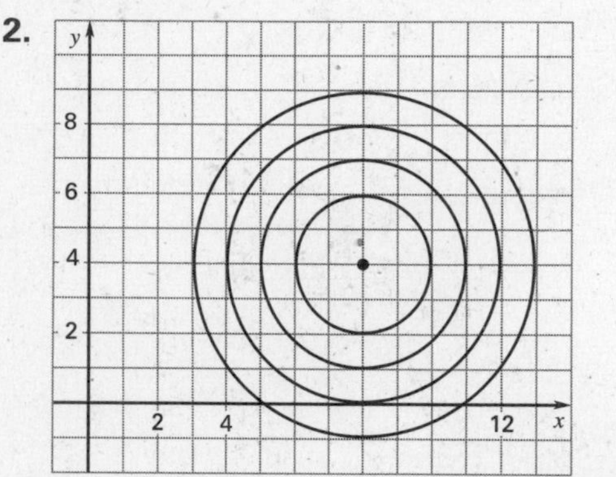

3. (7, 4): 12 points; (10, 7): 3 points; (10, 0): 1 point; (9, 6): 9 points; so total point value for turn = 25 points

4. $1200 \times 4 = 4800$ balls were rolled down the ramp; $5\% \times 4800 = 240$ balls landed in 12-point circle; $15\% \times 4800 = 720$ balls in 9-point circle; $30\% \times 4800 = 1440$ balls in 3-point circle; $50\% \times 4800 = 2400$ balls in 1-point circle

Challenge: Skills and Applications

1. a. $-14, 10$ **b.** 4
c. $(x+2)^2+(y+4)^2=16$, $(x+2)^2+(y+4)^2=484$

2. *Sample answer:* The midpoint of the chord $\overline{AB}$ is $M=\left(\frac{r+s}{2},\frac{t}{2}\right)$. The slope of $\overline{AB}$ is $\frac{t}{s-r}$, so the slope of the perpendicular bisector is $-\frac{s-r}{t}=\frac{r-s}{t}$. Therefore, an equation of the perpendicular bisector is

$y-\frac{t}{2}=\frac{r-s}{t}\left(x-\frac{r+s}{2}\right)$, or

$y=\frac{r-s}{t}x+\frac{t}{2}-\frac{r^2-s^2}{2t}$. But (s, t) is a point on the circle $x^2+y^2=r^2$, so $s^2+t^2=r^2$ and $\frac{r^2-s^2}{2t}=\frac{t^2}{2t}=\frac{t}{2}$. So the equation of the perpendicular bisector simplifies to $y=\frac{r-s}{t}x$. Since the center of the circle, (0, 0), satisfies the equation, the center lies on the perpendicular bisector of the chord.

3. *Sample answer:* First, observe that $B(s, t)$ is a point on the circle $x^2+y^2=r^2$, so $s^2+t^2=r^2$. Therefore,

$(AB)^2+(BC)^2=(s-r)^2+t^2+(s+r)^2+t^2$
$=(s^2-2sr+r^2)+t^2+(s^2+2sr+r^2)+t^2$
$=2r^2+2(s^2+t^2)$
$=2r^2+2r^2$
$=4r^2=(2r)^2=(AC)^2$.

Since $(AB)^2+(BC)^2=(AC)^2$, $\triangle ABC$ is a right triangle by the Converse of the Pythagorean Theorem.

4. $(-2, -3)$; 7 **5.** $(5, -4)$; 8 **6.** $(-1, 0)$; 6
7. $(-3, 4)$; 5 **8.** $(-3, 7)$; $\sqrt{70}$ **9.** $(4, 2)$; $\sqrt{2}$

Lesson 10.7

Warm-Up Exercises

1. point **2.** parallel **3.** center **4.** interior

Daily Homework Quiz

1. $(-3, 5)$, 6 **2.** $(4, -1)$, 7
3. $(x-1)^2+y^2=4$
4. $(x-3)^2+(y+2)^2=36$

Lesson Opener

Allow 20 minutes.

1. Check art. The set of points is a circle with center C and radius 8 cm. **2.** Check art. The set of points is the interior of a circle with center C and radius 8 cm. **3.** Check art. The set of points is two lines, each parallel to k and 2 inches from k. **4.** Check art. The set of points is a line that is parallel to m and n and halfway between m and n.
5. Check art. The set of points is the perpendicular bisector of $\overline{AB}$.

Lesson 10.7 *continued*

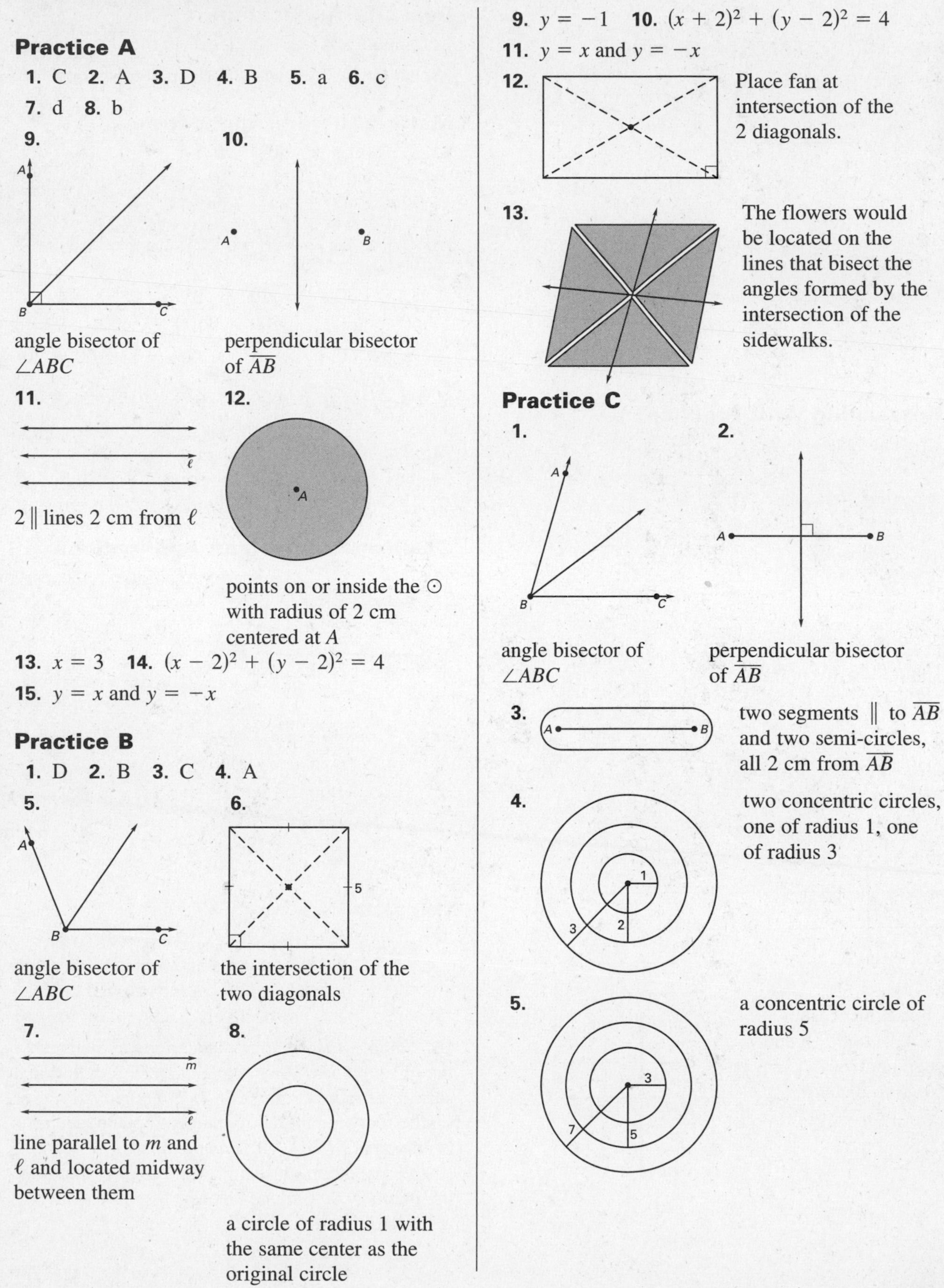

Practice A

1. C **2.** A **3.** D **4.** B **5.** a **6.** c **7.** d **8.** b

9. angle bisector of $\angle ABC$

10. perpendicular bisector of $\overline{AB}$

11. 2 ∥ lines 2 cm from ℓ

12. points on or inside the ⊙ with radius of 2 cm centered at A

13. $x = 3$ **14.** $(x - 2)^2 + (y - 2)^2 = 4$

15. $y = x$ and $y = -x$

Practice B

1. D **2.** B **3.** C **4.** A

5. angle bisector of $\angle ABC$

6. the intersection of the two diagonals

7. line parallel to m and ℓ and located midway between them

8. a circle of radius 1 with the same center as the original circle

9. $y = -1$ **10.** $(x + 2)^2 + (y - 2)^2 = 4$

11. $y = x$ and $y = -x$

12. Place fan at intersection of the 2 diagonals.

13. The flowers would be located on the lines that bisect the angles formed by the intersection of the sidewalks.

Practice C

1. angle bisector of $\angle ABC$

2. perpendicular bisector of $\overline{AB}$

3. two segments ∥ to $\overline{AB}$ and two semi-circles, all 2 cm from $\overline{AB}$

4. two concentric circles, one of radius 1, one of radius 3

5. a concentric circle of radius 5

Lesson 10.7 *continued*

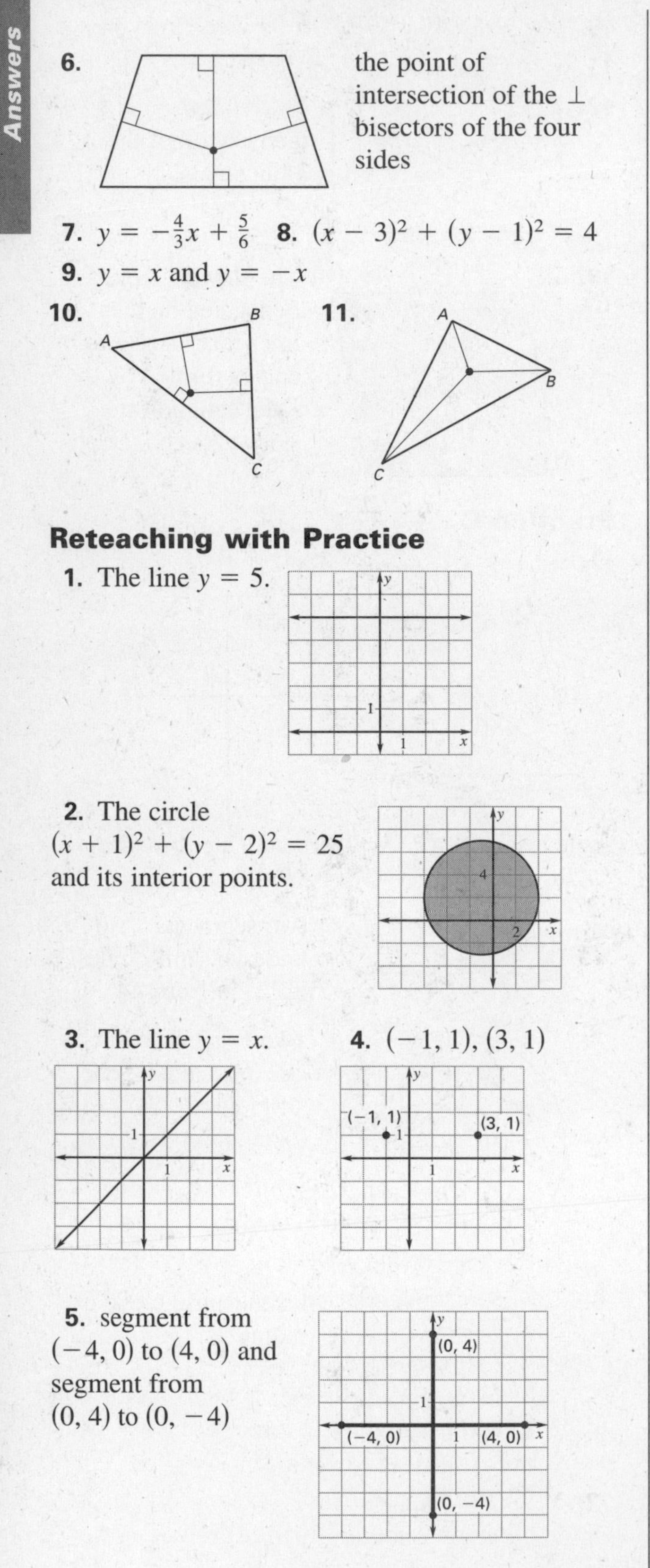

6. the point of intersection of the ⊥ bisectors of the four sides

7. $y = -\frac{4}{3}x + \frac{5}{6}$ **8.** $(x - 3)^2 + (y - 1)^2 = 4$

9. $y = x$ and $y = -x$

10.

11.

Reteaching with Practice

1. The line $y = 5$.

2. The circle $(x + 1)^2 + (y - 2)^2 = 25$ and its interior points.

3. The line $y = x$.

4. $(-1, 1), (3, 1)$

5. segment from $(-4, 0)$ to $(4, 0)$ and segment from $(0, 4)$ to $(0, -4)$

Real-Life Application

1. about 22.4 feet **2.** 15 or 20 foot knot **3.** about 7.4 feet **4.** Check student's work.

Math & History Application

1. $2 \cdot 60^2 + 1 \cdot 60 + 36$
$= 2 \cdot 3600 + 1 \cdot 60 + 36$
$= 7200 + 60 + 36 = 7296;$

$$22\left(\frac{1}{60}\right) + 30\left(\frac{1}{60}\right)^2 = 22\left(\frac{1}{60}\right) + 30\left(\frac{1}{3600}\right)$$

$$= \frac{22}{60} + \frac{30}{3600} = \frac{1320}{3600} + \frac{30}{3600} = \frac{1350}{3600} = \frac{3}{8}$$

2. a. $9 \cdot 60 + 48$ **b.** $1 \cdot 60^2 + 13 \cdot 60 + 26$ **c.** $45\left(\frac{1}{60}\right)$ **3. a.** 73,304 **b.** $= 3.125 = 3\frac{1}{8} \approx \pi$

4. *Sample answer:* 60 is a relatively small number, but it has many divisors. 1, 2, 3, 4, 5, 6, 10, 12, 15, 20, and 30 all divide into 60 evenly.

Challenge: Skills and Applications

1. **2.** no points

the incenter of the circle

3.

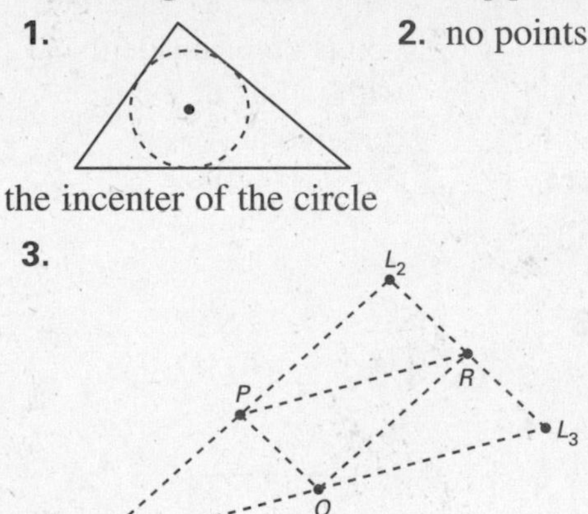

three points, as shown

4.

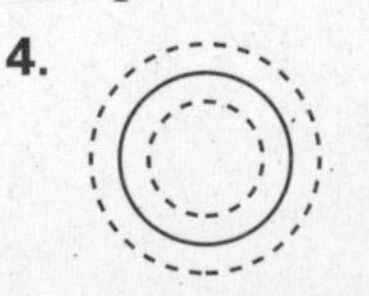

two circles, whose radii are 5 mm less and 5 mm greater than the radius of the original circle, respectively

5. *Sample answer:* one plane, midway between the given planes **6.** *Sample answer:* a cylindrical surface of radius 3 inches **7.** *Sample answer:* two planes that are parallel to the given plane, each 6 centimeters from it **8.** *Sample answer:* a plane (the perpendicular bisector of the segment joining the points) **9.** *Sample answer:* a sphere centered at the origin with radius 7

Review and Assessment

Review and Assessment

Review Games and Activities

1. 10 **2.** 2 **3.** 1 **4.** 8 **5.** 3 **6.** 9°
7. 6 **8.** 6° **9.** 5° **10.** 5° **11.** 4 **12.** 7

Diagram 1

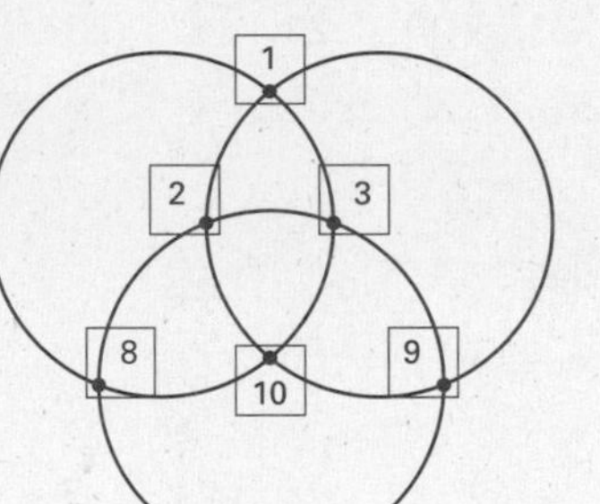

Diagram 2

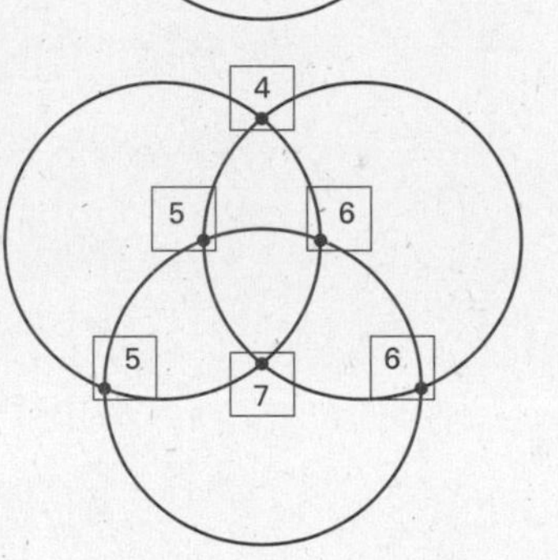

Test A

1. 4 ft **2.** 4.5 cm **3.** 1.05 m **4.** 42 cm
5. 66 ft **6.** 5.8 m **7.** G **8.** D **9.** B **10.** F
11. H **12.** A **13.** E **14.** C **15.** 60°
16. 55° **17.** 55° **18.** 125° **19.** 115°
20. 235° **21.** 180° **22.** 42° **23.** 130°
24. 100° **25.** 130° **26.** 270° **27.** 160°
28. 65° **29.** 29° **30.** 40° **31.** $\frac{17}{8}$ **32.** 5
33. $(-3, 4)$; 6 **34.** $(-1, 6)$; 1.5 **35.** A circle with a radius of 3 in. and a center at point C

Test B

1. 7.5 ft **2.** 5.75 cm **3.** 12.625 in.
4. 21 cm **5.** 200 in. **6.** 15.5 m **7.** E **8.** H
9. C **10.** F **11.** D **12.** A **13.** G **14.** B
15. 55° **16.** 35° **17.** 145° **18.** 125°
19. 90° **20.** 125° **21.** 90° **22.** 94° **23.** 60°
24. 250° **25.** 98° **26.** 118° **27.** 90°
28. 59° **29.** 35° **30.** 17.5° **31.** 2 **32.** 4
33. $(-2, -2)$; 5 **34.** $\left(-\frac{1}{2}, 6\right)$; $\sqrt{35}$
35. A circle with a radius of 5 cm and a center at point D

Test C

1. 7.75 ft **2.** 55 in. **3.** 2.5 m **4.** 50 cm
5. 19.5 ft **6.** 2.25 m **7.** D **8.** H **9.** A
10. F **11.** B **12.** E **13.** G **14.** C **15.** 75°
16. 15° **17.** 90° **18.** 105° **19.** 195°
20. 255° **21.** 40° **22.** 180° **23.** 79.5°
24. 120° **25.** 106° **26.** 180° **27.** 124°
28. 50° **29.** 105° **30.** 54° **31.** 5 **32.** $\frac{\sqrt{5}}{4}$
33. A circle with a radius of 7.2 cm and center at point A

SAT/ACT Chapter Test

1. B **2.** D **3.** A **4.** B **5.** E **6.** D **7.** C
8. D **9.** E **10.** A

Alternative Assessment

1. Complete answers should include:

Measure of the angle formed by the tangent and $\overline{AC} = 90°$

2. a. 114°; minor **b.** 246°; major **c.** 57°
d. $x^2 + y^2 = 144$ **e.** $(x - 5)^2 + (y + 7)^2 = 144$

3. a. 10 **b.** 24 feet **4.** Answers should include the three following cases: on the circle, inside the circle, and outside the circle.

Project: Proving a Conjecture

1. Yes; *EFGH* appears to be a square; *Sample answer:* The points E, F, G, and H are the same points as the vertices of the original square.

2. Yes; *PQRS* appears to be a rectangle. *Sample answer:* I did not notice anything unusual.

3. It is not possible to inscribe a nonrectangular parallelogram in a circle because opposite angles of a nonrectangular parallelogram will never be

Review and Assessment *continued*

supplementary. It is not possible to circumscribe a circle around a nonrectangular parallelogram because the circle determined by three points of the parallelogram does not contain the fourth point. **4.** *Sample construction:* A scalene quadrilateral $ABCD$ is inscribed in a circle. The points where the angle bisectors of $ABCD$ meet the circle are labeled Q, R, S, and T (where Q is the bisector of $\angle A$, R is the bisector of $\angle B$, and so on). *Sample conjecture:* The resulting quadrilateral $QRST$ is a rectangle.

5. *Sample proof:*
$m\overset{\frown}{STQ} = m\overset{\frown}{SB} + m\overset{\frown}{BQ} = 2m\angle BCS + 2m\angle BAQ = m\angle BCD + m\angle BAD = 180°$; $m\overset{\frown}{STQ} = 180°$, so $m\angle SRQ = 90°$. Then $m\angle STQ = 90°$ because a quadrilateral can be inscribed in a circle if and only if its opposite angles are supplementary. Use similar steps to prove that the other two angles of $QRST$ are right angles. Since $QRST$ is a quadrilateral with four right angles, it is a rectangle.

Cumulative Review

1. $m\angle A = 40°, m\angle B = 50°$ **2.** $m\angle A = 71°, m\angle B = 19°$

3.

Statements	Reasons
1. $\overline{AC} \cong \overline{BD}$	1. Given
2. $\overline{BC} \cong \overline{BC}$	2. Reflexive property
3. $\overline{AB} \cong \overline{CD}$	3. Subtraction prop. of equality

4.

Statements	Reasons
1. $\angle 1$ and $\angle 2$ are a linear pair.	1. Given
2. $\angle 1$ and $\angle 2$ are supplementary.	2. Linear Pair Postulate
3. $m\angle 1 + m\angle 2 = 180°$	3. Definition of supplementary $\angle$s
4. $\angle 2 \cong \angle 3$	4. Given
5. $m\angle 2 = m\angle 3$	5. Def. of $\cong$ $\angle$s
6. $m\angle 1 + m\angle 3 = 180°$	6. Substitution prop. of equality

5. $y = -2x + 3$ **6.** $y = 7x - 1$ **7.** $m\angle T = 55°, m\angle S = 70°$ **8.** $m\angle J = 70°, m\angle L = 70°$ **9.** 17 **10.** 15 **11.** $x = 40, y = 3$ **12.** $x = 1, y = 5$ **13.** $(-1, 11)$ **14.** 8 **15.** 120° **16.** 12 **17.** 30 **18.** 2 **19.** $\frac{10}{3}$ **20.** 150° **21.** 100° **22.** 48° **23.** 160° **24.** 28° **25.** 122° **26.** $x = 12, y = 8$ **27.** $x = 6, y = 9$ **28.** $(x - 2)^2 + (y - 3)^2 = 16$ **29.** $(x + 1)^2 + (y - 2)^2 = 9$